Introduction to
Petroleum
Exploration and Engineering

Introduction to
Petroleum
Exploration and Engineering

Andrew Palmer

W **World Scientific**

NEW JERSEY · LONDON · SINGAPORE · BEIJING · SHANGHAI · HONG KONG · TAIPEI · CHENNAI · TOKYO

Published by

World Scientific Publishing Co. Pte. Ltd.

5 Toh Tuck Link, Singapore 596224

USA office: 27 Warren Street, Suite 401-402, Hackensack, NJ 07601

UK office: 57 Shelton Street, Covent Garden, London WC2H 9HE

Library of Congress Cataloging-in-Publication Data
Names: Palmer, Andrew C. (Andrew Clennel), 1938–   author.
Title: Introduction to petroleum exploration and engineering / Andrew Palmer, NUS, Singapore.
Description: New Jersey : World Scientific, [2016] | Includes bibliographical references.
Identifiers: LCCN 2016028684| ISBN 9789813147775 (hardback : alk. paper) |
        ISBN 9789813147782 (pbk. : alk. paper)
Subjects: LCSH: Petroleum engineering. | Petroleum--Prospecting. | Natural gas.
Classification: LCC TN870 .P25 2016 | DDC 622/.3382--dc23
LC record available at https://lccn.loc.gov/2016028684

British Library Cataloguing-in-Publication Data
A catalogue record for this book is available from the British Library.

Desk Editor: Amanda Yun

Typeset by Stallion Press
Email: enquiries@stallionpress.com

Foreword

Petroleum supplies most of the energy that keeps civilisation going. That is almost certain to remain the case through our lifetimes and for some time beyond, though in the very long term petroleum will be used up — but hopefully not before we find a plentiful replacement that will resolve our energy and environmental debates forever. In the meantime, the petroleum industry will continue to thrive.

The petroleum industry is a challenging, enjoyable, and lively field to study and work in. Andrew Palmer has written this most interesting book for newcomers to the industry, assuming no prior knowledge and based on his experience of teaching at Cambridge, Harvard and the National University of Singapore. I have no doubt that readers of all ages will find the subject as exciting as we do in practice.

Enjoy!
Hamid D. Jafar
Chairman, Crescent Petroleum
Sharjah, 7 June 2016

Contents

About the Author

Andrew Palmer is retired Professor of the Center for Offshore Research and Engineering of the Department of Civil & Environmental Engineering of the National University of Singapore (2006–2015), as well as retired Jafar Research Professor of Petroleum Engineering, Cambridge University, UK. He is a Fellow of the Royal Society, a Fellow of the Royal Academy of Engineering, a Fellow of the Institution of Civil Engineers, and a Chartered Engineer in the UK.

He has divided his career equally between practice as a consulting engineer and university teaching. He was appointed as a Lecturer in the Department of Mechanical Engineering of the University of Liverpool in 1965, and in 1968 as a University Lecturer in the Department of Engineering in the University of Cambridge. In 1975 he joined R.J. Brown and Associates, a company of consulting engineers based in the Netherlands, and was appointed Chief Pipeline Engineer in 1976. In 1979 he was appointed as Professor of Mechanical Engineering in the Department of Mechanical Engineering of the University of Manchester Institute of Science and Technology. In 1982 he returned to R.J. Brown and Associates as Vice-President Engineering, responsible for technical development and innovation.

In 1985 he founded Andrew Palmer and Associates, a company of consulting engineers principally engaged with marine pipelines, but also in defence and Arctic engineering. The company grew to 55 people in offices in London, Glasgow and Aberdeen. The company was bought by Science Applications International Corporation in 1993, and later bought by Penspen.

In 1996 he returned to research and university teaching as Research Professor of Petroleum Engineering at Cambridge University in England. He was a Visiting Professor in the Division of Engineering and Applied Sciences at Harvard University, 2002–03. In 2005 he retired from Cambridge and moved to the National University of Singapore, where he is active in research, graduate education and management. He retired from NUS in 2015. He has an active consulting practice, much of it concerned with legal disputes and accident investigation.

He is the author of five books and more than 270 papers on pipelines, offshore engineering, geotechnics, ice and Arctic engineering. He was Chairman of the DNV Pipelines Committee from 2007 until 2010.

Andrew Palmer is married to Jane, and they have one daughter, Emily.

Introduction

The inspiration for this book comes from my teaching at Cambridge University, Harvard University, and the National University of Singapore. A second inspiration has been my work in the petroleum industry, which I was lucky enough to get involved with forty-five years ago.

I start off my courses in petroleum engineering by asking each student what she or he knows about the subject already. The questionnaire does not count for marks and can be submitted anonymously if the student prefers. In most cases, a student knows very little: why should he or she? He does not know where petroleum comes from, or what it consists of, or how old it is. Still less does he know about concepts like permeability, the Jurassic era and left-lateral faults, or industry language like kelly, stinger, mousehole and possum belly tank. It will go without saying that we must not reproach a student for those gaps in knowledge, which are perfectly understandable, but it does indicate that there is a need for an introductory text that assumes almost no prior knowledge, at least not in petroleum.

This is that introduction. It assumes that a reader knows nothing at all about petroleum. In a few places, it does assume a little knowledge of high-school physics and mathematics — for example, that gas is compressible and what a derivative is. The reader who does not want to pursue those topics can skip the relevant sections.

The industry or academic reader should keep in mind that the book was not written primarily for her or him, but for a student with a far more restricted prior knowledge. If she thinks that the text is very simple, I shall regard that as a compliment. If she thinks that a subject has been so far simplified as to become seriously misleading, that is of course important: I apologise and would like her to tell me why. Every reader — even including the writer — will think that his pet subject has been given too little attention. There are excellent specialised books and papers about every aspect of the subject, and I refer to many of them.

I am grateful to many colleagues, students and friends for their insight, encouragement and inspiration, among them Bob Brown, Chan Eng Soon, Choo Chiao Beng, Choo Yoo Sang, Ann Dowling, the late Jack Ells, Simon Falser, Ian Fitzsimmons, Mike Gibson, John Halkyard, Herbert Huppert, Hamid Jafar, John Kenny, Matilda Loh, Bryan Lovell, Dan McKenzie, Kristina Moreno, Amanda Pyatt, the late Allan Reece, Ekhard Salje, Zuraidah Sapari, Andrew Schofield, the late Ted Schultz, Tan Thiam Soon, John Thorogood, Hendrik Tjiawi, Too Jun Lin, David Walker, Ron Watkins, Wong Sau Wei, Andrew Woods, Xie Peng and Zheng Jiexin. I should also like to thank Amanda Yun and her colleagues at World Scientific for encouraging me to write and finish the book, and for their care in its preparation. I am grateful to my wife Jane for her infinite patience and forbearance.

The mistakes and misjudgments are mine, and I would like to know about them.

My mentor Allan Reece used to say that nothing was worth spending time on if it was not enjoyable. It may be impracticable to stick to that rigorously in the real world, but he had the right idea. I hope you will enjoy this book.

Andrew Palmer

Singapore

February 2016

Chapter One

Petroleum and Human Society

1.1 Introduction

Human society has come to rely on the availability of enormous amounts of energy. That energy helps grow our food, powers our factories to make all the things we think we need, gives us light at night, keeps us warm in cold climates (and cool in hot climates), makes it possible for us to communicate information, and helps us move about the surface of the earth. Without that energy, and without the energy being available at a tolerable cost, life as we know it would be impossible.

Our civilisation exploits fossil fuels in huge quantities. World primary energy consumption in 2014 was 12.9 Gtonnes oil equivalent (BP, 2015), when all forms of energy are converted to equivalent oil; 1 Gtonne is 1 thousand million metric tonnes. One metric tonne of oil is equivalent to 42 GJ; 1 GJ is a measure of energy, and is 3.6×10^9 kW hour. In a few countries consumption has stopped increasing and levelled out, but in most countries, and particularly in those that were in the past described as 'underdeveloped', consumption continues to increase inexorably, over the whole world from 8 Gtonnes a year 25 years ago.

Overwhelmingly, that energy comes from fossil fuels, which are oil, natural gas and coal. Oil and natural gas together account for 56% of total primary energy, and coal for 30%. Energy-related consumption dumped 31.6 Gt of carbon dioxide into the atmosphere in 2012, and the scientific consensus is that it is leading to climate change, though there are dissenters. Alternative energy sources are examined in section 1.2, but at present their contribution to humankind's energy needs is relatively small. That is undeniably true today, when renewables contributed only 2.5% of primary energy.

A more controversial question is whether it will still be true in fifty years' time, towards the end of the lifetime of anyone reading this.

It is notoriously difficult to try to predict the future, but in the author's opinion, humankind will depend on fossil fuels to roughly the same extent in 2065 as it does in 2015. To say this is not to express approval but simply to dispassionately summarise what seems most likely to happen.

Unconventional gas and oil have scarcely been touched. Shale gas has only come into fashion in the last decade, but is already having major economic and political effects. Huge amounts of gas are locked up in gas hydrates, which are solid compounds of gas and water stable at low temperatures and high pressure: they have not yet been begun to be exploited. One estimate said that the gas in known fields of hydrates corresponds to 700 years of current gas consumption, though that estimate is not linked to an analysis of how much it will cost to produce that gas.

Humankind has been using fossil fuels for a long time. For two thousand years China has burned natural gas in Sichuan to evaporate brine to make salt (Zheng and Palmer, 2009). The ancient Babylonians used natural petroleum from seeps, both as a fuel and as a sealant (as recounted in the Old Testament). Fossil fuel consumption remained small until it began to take off some two hundred years ago. The modern development of coal is recent, and of oil and gas still more recent.

In the long term, that exploitation of naturally-formed fossil fuels must clearly be unsustainable. Humankind is enthusiastically and heedlessly using up coal and petroleum that took hundreds of millions of years to form, and by our undeserved good luck has remained trapped over all that time. Those resources are not being replaced, and they cannot be replaced. Things have to change. There is no prospect whatsoever that humankind will learn to live without energy, and so at some point in the future there will have to be an immense effort to replace fossil fuels with another source. That would still be true even if there were not the carbon dioxide pollution issue discussed in section 1.3 below.

1.2 Alternatives to Petroleum

Eighty percent of world energy needs are met by fossil fuels, and about half of that by coal and the other half by oil and gas. As illustrations of the scale of the issue, global coal consumption grew from 4.8 Gtonnes in 2000 to 7.7 Gtonnes in 2012, and in the seven years up to 2012 China installed 150 MW of electricity generation capacity every day, mostly based on coal. Alternatives to petroleum are a huge subject, and this book is not intended to be about alternatives. For the several reasons set out in Section 1.1, the current age of petroleum cannot continue indefinitely. That is scarcely an immediate problem, but it is instructive to think about how humanity will secure the energy it needs after petroleum is gone. There is an extensive literature, and the International Energy Agency website (www.iea.org) is a good and balanced starting point.

One source of energy is nuclear. The potential for nuclear energy was recognised in the late 1930s, and the first atomic bombs were detonated in 1945, initially as a demonstration in July and later in the destruction of Hiroshima and Nagasaki. The first nuclear power stations went into service in 1954, and there are now more than 400 nuclear power plants, which together generate about 15% of humanity's electricity. The development is grossly uneven, reflecting conflicting policies in different countries, prompted by conflicting evaluations of the risk of a nuclear accident like Fukushima in 2010 and Chernobyl in 1976, as well as by political pressures. On the one hand, France generates some 80% of its electricity from nuclear, while China is proceeding vigorously and starts up a new nuclear plant every six months, but on the other hand Australia has no nuclear power and other countries such as Germany have said that they will close down the nuclear power they currently have (though it remains to be seen if it will follow through, and there is significant opposition to the change). Many more countries are contemplating nuclear power. Palmer *et al.* (2010, 2015) examine the special case of Singapore.

Existing nuclear power plants are based on fission reactions based on uranium. Fission is a reaction when the nucleus of an atom, having captured a neutron, splits into two or more nuclei, a process that releases more neutrons and a significant amount of energy. Those neutrons go on to split more nuclei and a chain reaction takes place. Uranium is found quite widely in different regions of the Earth, but the supply is not unlimited. One alternative is to base the nuclear fuel cycle on fission in thorium, which is more widely distributed than uranium. In the long term, though, a better choice is likely to be to apply the nuclear fusion reaction that fuels the hydrogen bomb. Fusion is a process where nuclei collide and join together to form a heavier atom, usually deuterium or tritium, in the process releasing a lot of energy at an extremely high temperature. In the bomb context, that possibility was first recognised in the late 1940s, and the first bomb was exploded in 1952. Much work has been done to try to tame the hydrogen fusion reaction so that it can become a continuous power source. Some sceptics argue that application of the hydrogen reaction is nowadays often said to be fifty years away in the future, and that it will always be fifty years away, but others are more optimistic. Once the hydrogen reaction has been tamed, humanity will never again have an energy problem. An interim option that has been advanced (Hawthorne, 1970) is repeatedly to explode hydrogen bombs underground and to recover the released heat to power a conventional thermal power system.

'Renewables' provided 19% of electricity and 13.1% of primary energy supply in 2009, though only 3% of energy for transport. Conventionally, nuclear energy is not counted as a renewable.

Hydropower uses the potential energy of water at a height to drive a turbine and a generator. It was the earliest form of renewable energy to be used extensively, and is the world's largest source of renewable electricity. The capital cost is often high, and the environmental impact, loss of agricultural land, and disruption to wildlife are all significant. Many of the best sites have been developed, but many possibilities remain, notably the enormous Lena, Yenisei and Ob' rivers in Siberia.

Geothermal energy uses the temperature difference between the surface and hot rocks deep into the earth to drive a heat engine. It usually requires drilling to 2 or 3 km, but less in volcanic regions where the temperature gradient is unusually large.

Ocean thermal energy (OTEC) uses the temperature difference between warm water at the ocean surface and cold water at a depth of 1 or 2 km in some locations in the tropics. It has been shown to work on a very small scale. If it is possibly to be economic, it has to have a good site with a temperature difference of 20° or so, and it has to be on a large scale: one typical scheme required a cold water pipe 7 m in diameter.

Wind energy is another alternative. Onshore wind is attractive to politicians, and has been widely developed in a few countries, though the visual, environmental and noise impacts are significant, costs are high, and most projects seem to require some kind of artificial subsidy. In most locations, the wind does not blow strongly and consistently all the time, and windmills produce no power when the wind is light and have to be shut down when the wind is high. Offshore wind has much higher costs, both for construction and for operation, but is being pursued, for example in the North Sea. The penetration of the energy market by wind energy is for the moment small. Intermittent supply is a significant difficulty: the system generates nothing if there is no wind, but it has to be rugged enough to withstand occasional very high winds. Taking a lot of energy out of the wind would itself have a significant effect on climate.

In contrast, tidal power is wholly predictable. It has been exploited on a small scale for a long time in a few locations. Rance in northern France has a tidal power plant that generates 240 MW at its peak, but on average produces 62MW; and a slightly larger plant has gone into service at Sihwa Lake in South Korea. Most tidal plants are very much smaller. There are other potential sites in macrotidal areas such as the Bay of Fundy in Canada, the Bristol Channel in England and the Sea of Okhotsk in Russia. It has been estimated that a Bristol Channel tidal power installation could generate 9% of UK's electricity. The capital cost is high, and the environmental impact is substantial.

The most commonly applied version of solar power depends on power cells that take in light energy and convert it to electricity. They do not work at night, and their output is much reduced on cloudy days. The energy generated in the daytime cannot easily be stored. The cost is high, but it has fallen substantially because of growing demand and increased production volumes, and it is now within sight of the level at which it is competitive with fossil fuels. Solar is the fastest growing renewable power technology worldwide, and installed capacity reached 67 GW at the end of 2011. An alternative is concentrated solar power, which uses a system of mirrors to reflect sunlight onto a boiler, but that option is applied much less often.

Yet another alternative is to grown fuel crops such as cane sugar, beet sugar, palm oil and corn, to convert them into liquid fuels by a fermentation process, and then to burn the fuels in thermal power stations or to power vehicles. This is being done in Brazil and the USA, and in Brazil biofuels from sugar cane accounts for 23% of transport fuel. Fuel crops can be thought of as one way of exploiting solar energy, and of avoiding the difficulty that sunshine is only in the daytime and not continuous. The crops take in carbon dioxide as they grow, and the carbon is released back into the atmosphere as carbon dioxide when the fuel is burned, so the overall process is carbon-neutral. A large area of land is needed, and there is an unfortunate competition with food crops.

Instead of conventional fuel crops, an alternative is to use the sunshine to grow algae in water, recover fats from the algae, and make the fats into fuel. The capital and operating costs seem likely to be very high indeed.

All these schemes are being studied, and each of them has enthusiasts. An indications of their relative popularity is given by International Renewable Energy Authority (IRENA) numbers about the numbers of people employed in different areas. Some 6.5 million jobs depend directly or indirectly on renewable energy. Solar energy has passed biofuels as the leading generator of employment, with 2.3 million solar jobs in 2013 against 1.5 million in biofuels. A lot of heavy labour is required by biofuels, but that may be an advantage because it creates opportunities for unskilled people with limited education. In contrast, uncertainty about the future development of wind power in the US, Europe and India has limited employment to 0.8 million jobs, though China and Canada are more positive.

It is completely clear that there is no easy answer. Many claims for renewable energy are made, and generous subsidies have been available for any research project that appears even marginally credible. The political establishment wishes to be able to deflect criticism by saying that it is supporting research: it can thereby avoid taking any significant action. Few of the claims for renewable energy sources are the object of constructive scepticism, particularly about costs, and there is no subsidy for anyone who would like to test the claims.

For the present, humanity will continue to rely on petroleum and other fossil fuels.

1.3 Climate

Burning hydrocarbons releases carbon dioxide into the air. The carbon dioxide content of the atmosphere has increased from about 270 ppm (parts per million) in the pre-industrial age before 1750 to more than 400 ppm today, in the way shown in Figure 1.1. The rate of increase shows no sign of abating. The atmosphere is well-mixed, so the increase is almost the same everywhere on the Earth. As climate scientists have been pointing out for more than seventy years:

> 'We are carrying out a grand experiment on the atmosphere, and hoping for a null result.'

There is strong scientific evidence that this increase in carbon dioxide content is leading to climate change and to significant warming. Fourier recognised in 1827 that carbon dioxide and other gases in the atmosphere create a natural greenhouse effect, so that the Earth's surface is on average about 21°C warmer than it would be if there were no atmosphere. In 1896 the Swedish chemist Svante Arrhenius and geologist Arvid Högbom (Crawford, 1996) pointed out that the climate might be changed by increasing the amount of carbon dioxide in the atmosphere and adding to the natural effect. They thought global warming to be a good thing, which is understandable if one lives in Sweden, and it was proposed that shallow coal seams be exposed and deliberately burned to enhance the warming. Houghton (1994) describes the mechanism. Briefly, atmospheric carbon

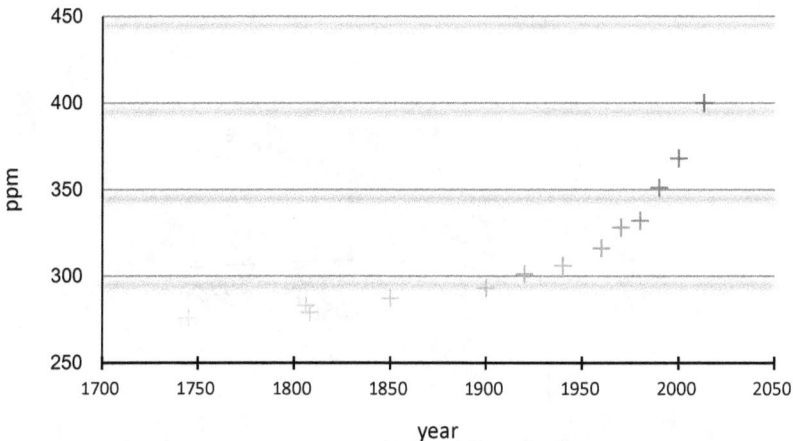

Figure 1.1   Carbon dioxide concentration in the atmosphere

dioxide absorbs some of the energy radiated from the Earth, so that less energy is radiated into space. The temperature rises until the Earth comes back into balance and the outgoing energy matches the incoming solar energy. If there were no additional feedbacks, doubling carbon dioxide in the atmosphere would lead to a 1.2°C increase of surface temperature, but feedbacks increase that to 2.7°C. There are many complications and subtleties, and Arrhenius took account of some of them.

Pacala and Socolow (2004) of Princeton University put together some thought-provoking numbers. Figure 1.2 illustrates the scale of the issue: it plots carbon emissions into the atmosphere, starting in 2000 when they were about 7 Gt/year and by 2050 climbing to 14 Gt/year, assuming a business-as-usual strategy where development continues and no significant action is taken to restrict carbon emissions.

Some of those emissions are from the combustion of coal. Figure 1.3 plots fossil fuel consumption, and shows dramatic increases, most of all in China.

The current carbon dioxide content of the atmosphere is 400 ppm (parts per million), a long way up from the pre-industrial level of 270 ppm three hundred years ago. The atmosphere is well-mixed, and the consensus opinion is that 400 ppm has already caused a change in climate and temperature. The atmosphere is well-mixed, but the changes are not quite uniformly distributed. It ought to be added that not everyone agrees, and that there is substantial disagreement about the scale of the temperature

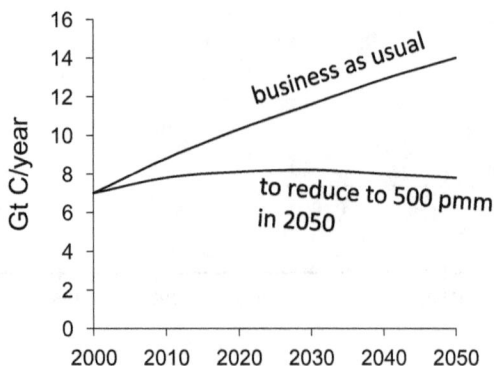

Figure 1.2   Carbon emissions

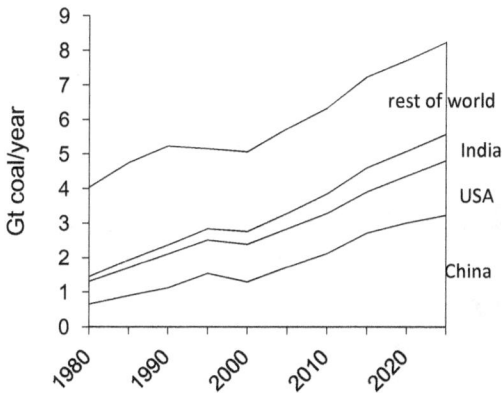

Figure 1.3   Forecast world consumption of fossil fuel

increase. Some optimists think that it may be possible to keep atmospheric carbon dioxide below 500 ppm, and Figure 1.2 shows that to achieve that target emissions would have to level out at around 7 Gt/year, starting in 2004. Pacala and Sokolow focussed minds by asking what would have to be done to reach that reduction. They introduced the concept they called a "wedge": one wedge is a reduction of emissions that starts at zero and increases linearly with time to 1 Gt/year in 2050, so seven wedges would be required to reduce emissions to 7 Gt/year from the business-as-usual 14 Gt/year. They argued that

> 'Humanity can solve the carbon and climate problem in the first half of
> this century simply by scaling up what we already know how to do.'

and that represented a valuable corrective to the idea that the problem required some 'magic bullet' that depended on a technological break-through that has not been made.

Table 1.1 lists the actions that might be taken to save 1 wedge.

Time has passed since 2004, and almost nothing has been done. The general view is that we now need nine wedges rather than seven. Climate change denial continues, and every aspect of the discussion has become politicised. The argument has brought in people who can hardly be described as scientists, on one side in the USA Al Gore, formerly the Vice-President, and on the other side in the UK Nigel Lawson, formerly

Table 1.1   Actions that would reduce emissions by 1 Gt/year (1 "wedge")

action	scale required
switch fuel from coal to gas	1400 GW fuelled by gas instead of coal
switch from coal to nuclear	700 1 GW nuclear power stations (1.5 × current world nuclear capacity)
switch from coal to solar	2000 GW peak (700 × current world solar capacity, requiring 20,000 km² land area
switch from coal to wind	2×10^6 1 MW windmills (MORE × current world wind capacity)
switch from petroleum-based to biomass-based fuels	2.5×10^6 km² cropland (1/6 of world cropland)
carbon dioxide capture and storage	CO_2 from 700 1 GW coal-fueled power plants
less use of cars	decrease car travel from 16000 km/year to 8000 km/year for 2×10^9 cars

Chancellor of the Exchequer, and in the USA every Republican candidate for the upcoming Presidential election. In parallel, there is an ongoing argument that Pacala and Socolow focussed too much on technological fixes, and that there ought to have been a broader emphasis that included reduction in energy demand, by limiting population growth and turning back from energy-intensive lifestyles.

All the actions have very high costs, and they would almost all be controversial and politically unpopular. A little is happening. In rich countries, there is a limited move towards a switch from coal to gas as the fuel for electricity generation, principally because of environmental concerns but also because switching is a cheap option that generates political goodwill at little cost. In China there is an energetic nuclear energy programme, but nuclear remains a relatively small proportion of the rapidly-growing energy demand. In other parts of the world, nuclear development has been delayed or brought to a complete halt, in response to the 2010 Fukushima disaster. Solar is one of the more credible sources of alternative energy, but does not work at night and so has to be supplemented by another energy source. Wind energy is much in fashion, but it is expensive and fails if the wind is either too weak or too strong, and some estimates indicate that it is impracticable to operate a stable power grid if the wind component is more than 15%.

It may be doubted whether any government — whether democratic or undemocratic — would be ready to embark on most of those actions, at least not until the effects of climate change are seen to be much more serious than they are today. Unsurprisingly, little that is significant has been done so far, and emissions continue to grow.

A problem is that the atmosphere is well-mixed, and so the carbon dioxide content is nearly uniform. With most kinds of pollution, action taken to reduce pollution in one location has an immediate effect in that location and nearby, and therefore modest actions in one place can be seen to be beneficial. With carbon dioxide, on the other hand, the effects of action in one place are more or less evenly spread around the world, and therefore the local effect is limited.

The reality of the natural greenhouse effect is not disputed, but the argument that the effect of increased atmospheric carbon dioxide is a result of human activity is not accepted by all scientists. To make matters much worse, the issue has become deeply politicised. Some people regard acceptance or denial of the reality of human-induced global warming as a touchstone of political positions across a much broader range, and they abuse or shout down anyone who reaches a different conclusion. A full discussion of the issue is outside the scope of this book, but it is of concern to everyone who is engaged with the petroleum industry, because opinions about it — whether well-informed or ill-informed — influence responses to totally different questions. The reader will find statements of the widely accepted opinion in the reports of the Intergovernmental Panel on Climate Change (IPCC; see for example IPCC (1990 and 2007)) and Houghton (1994) and of the contrary opinion in many books and websites (see, for example, Carter (2010) and Montford (2010)). Statements that "the science is settled" and intended to shut off discussion need to be treated sceptically.

Many of those who accept that climate change is caused by human activity campaign vigorously for severe restrictions on the use of fossil fuels. Their efforts have so far had almost no effect. If one accepts the underlying argument, the question of what action to take is bitterly controversial and divisive. The countries that are now industrialising and using more and more fossil fuel argue that the problem was created in Europe and North America by the countries that industrialised earlier, and that those countries ought to bear most of the economic burden of making gross changes in energy sources. Those countries' scientists react by pointing out that the countries that are now developing rapidly are using so much fossil fuel that warming will proceed nearly unchecked unless everyone shares in the huge changes that will be needed. A pessimistic view is that little has been done, and that not much more will be done unless and until the effects of global warming are much more marked that they are today.

The issue is further influenced by changing feelings. Forty years ago, nuclear war, hunger and population growth were widely thought to be the overriding priorities. Those concerns have diminished but not disappeared, in part in response to technological change in agriculture, and in part because contraception and increases in prosperity in the less-developed world have slowed population growth. Current concerns centre on climate change, and many people argue that this is the most important problem and that humankind has to take drastic action to avoid it. That may be temporary. Human perceptions of urgency and priority might change rapidly, as could easily happen if there were a thermonuclear war or a continent-wide epidemic. Either could alter perceptions about priorities, in a matter of weeks.

If climate change is the worst calamity that befalls humanity in the next hundred years, we shall have been quite extraordinarily lucky.

References

BP Statistical review of world energy (2015).

Carter, R.M. Climate: the counter consensus. Stacey International (2010).

Crawford, E. Arrhenius: from ionic theory to the greenhouse effect. Science History Publications (1996).

Gold, T. The deep hot biosphere. Copernicus (1999).

Hawthorne, W.R. Personal communication (1970).

Houghton, J. Global warming: the complete briefing. Cambridge University Press, Cambridge (1994).

Hunt, J.M. Petroleum geochemistry and geology. W.H. Freeman (1996).

Intergovernmental Panel on Climate Change. Climate change: the IPCC scientific assessment. Cambridge University Press, Cambridge (1990).

Intergovernmental Panel on Climate Change. Climate change 2007: the physical science basis: contribution of working group 1 to the Fourth Assessment. Cambridge University Press, Cambridge (2007).

Mann, M.E. Dire predictions: understanding global warming. DK Publications, New York (2009).

Montford, A.W. The hockey stick illusion. Stacey International (2010).

Pacala, S. and Socolow, R. Stabilization wedges: solving the climate problem for the next 50 years with current technologies. *Science*, **305**(5686), 968–972 (2004).

Palmer, A.C., Ramakrishna, S. and Cheema, H.M. Nuclear power in Singapore. Institution of Engineers Singapore Journal Part A: Civil and Structural Engineering. Presented at Nuclear Power Asia 2010 Conference, Kuala Lumpur (2010).

Palmer, A.C. and Oliver, G.H.J. Institution of Engineers Singapore Journal Part A: Civil and Structural Engineering. Nuclear Power in Singapore (2015).

Zheng, J. and Palmer, A.C. Bamboo pipelines in ancient China (and now?). *Journal of Pipeline Engineering*, **8**, 95–98 (2009).

Chapter Two

What is Petroleum?

2.1 Introduction

This chapter sets out to explain what petroleum is. It takes many different forms, from gas lighter than air to thick bitumen like black and oily peanut butter. Petroleum is mostly composed of hydrocarbons, so the chapter begins by explaining what a hydrocarbon is. It goes on to consider the different fractions that petroleum can be divided into, starting with lightest (natural gas) and going on to the densest (bitumen).

Description of where petroleum comes from is kept back until Chapter 3.

2.2 Hydrocarbons

Petroleum consists mostly of hydrocarbons. A hydrocarbon is a compound of hydrogen atoms (denoted H) and carbon atoms (C). There are literally thousands of different hydrocarbons, distinguished from each other by both structure and the numbers of C and H atoms.

The numbers of atoms in a molecule are indicated by subscripts, so that C_2H_6 means that a molecule has two carbon atoms and six hydrogen atoms. C_2H_6 is ethane, the second in a series called alkanes. The subscript 1 is not used, so that CH_4 means that the molecule has one carbon atom and four hydrogen atoms. The convention is to write carbon C first.

Carbon has valency 4, which means that one carbon atom can form four links with other atoms, both with other carbon atoms and with atoms of other elements such as hydrogen, oxygen (O) and sulphur (S). That accounts for the astonishing variety and complexity of molecules that contain carbon. Many substances that include carbon are called organics, because at one time it was thought that they always derived from life, but that opinion was discredited when urea was synthesised from hydrogen, carbon, sulphur and nitrogen atoms that had sources nothing to do with life.

The most familiar compounds with other elements besides hydrogen are carbon dioxide (CO_2) and hydrogen sulphide (H_2S). Carbon dioxide is formed when materials that contain carbon are burned. Hydrogen sulphide is formed when organic materials that contain sulphur are allowed to rot, accounts for the smell of rotten eggs, and is poisonous if it can be smelled. It is deadly even at low concentrations: a 0.1% concentration is fatal in less than half an hour.

2.3 Natural Gas

Natural gas is mostly methane CH_4. One molecule consists of one carbon atom surrounded by four hydrogen atoms symmetrically placed. It should not be thought that the structure is rigid: the atoms can move relative to each other. Methane is the first in a family of hydrocarbons nowadays called alkanes (from the German for alcohol) but formerly paraffins (from Latin words that refer to their relatively low reactivity). Alkanes have the general formula C_nH_{2n+2} so that methane corresponds to *n* equal to 1.

Methane is a gas at ordinary temperatures, and it can be liquid only if it is colder that its boiling point, −161.5°C at atmospheric pressure, which is why tankers that carry natural gas have to have elaborate insulation and refrigeration systems.

Other hydrocarbons are present in much smaller proportions, less than 10% altogether. The second alkane is ethane C_2H_6 and the third propane C_3H_8. After that comes butane C_4H_{10}, but now there is an extra complication: butane can have two structures, both with the same chemical formula but with slight different physical properties, called isomers. Normal butane (*n*-butane) has the four carbon atoms in a single chain and −0.5°C boiling point, whereas isobutane has the four carbon atoms in a branched chain and −11.7°C boiling point. Propane and butane are the major constituents of 'bottled gas', used in camping stoves and where there is no mains gas supply. They are also a valuable feedstock to make other compounds. The number of possibilities increases more and more rapidly as n increases, so that there are two butanes, three pentanes, five hexanes, 18 octanes and 36,797,588 alkanes with 25 carbon atoms (Hunt, 1996).

Natural gas also contains other families of hydrocarbons, but in small fractions. Ethylene C_2H_4 is the first in a family with a general formula C_nH_{2n} (starting with n equal to 2), called alkenes (formerly olefins, from the French for oil-forming). Alkenes are more reactive than alkanes, because the carbon atoms are linked by two bonds rather than one, represented by two lines in a skeletal formula. Another atom can rather easily break one of the bonds and link to the two carbon atoms on either side. Again there can be branched chains and many isomers. Hydrocarbons can also have ring structures, starting with benzene C_6H_6, the first of a family with the general formula C_nH_{2n-6}, often known as aromatics because impurities give many of them a sweet smell, but now called arenes. A famous story tells how the chemist Kekulé was trying to puzzle out the structure of benzene, and in a dream saw a ring of snakes, each one biting the next one's tail. Some of the ring structures have several rings linked together.

As well as hydrocarbons, there can be other gases, such as nitrogen N_2, carbon dioxide CO_2 hydrogen sulphide H_2S and helium He. Usually their fractions are very small, but gas from the Sleipner field in the North Sea has 9% CO_2, and sometimes the fraction is higher still (Hunt, 1996). Gas with a significant amount of H_2S is called 'sour', which is important because it can be highly corrosive (Palmer and King, 2010) as well as toxic. Gas that is not sour is called 'sweet'.

2.4 Crude Oil

A crude oil is a mixture of many hydrocarbons and other compounds. Roughly half of a typical crude is composed of napthenes (cycloparaffins) which have five or six carbon atoms in a warped ring. There are many larger and more complex napthenes and aromatics with four or more conjoined rings. The next most important group of compounds are alkanes (paraffins), described in section 2.3. In small proportions, there are alkenes and nonhydrocarbons containing oxygen, sulphur and nitrogen. Many crude oils contain water and dissolved gas.

Almost every oil is different. One component of a description is API (American Petroleum Institute) gravity, a measure of density defined by

$$\text{API gravity} = \frac{141.5}{SG_{60}} - 131.5$$

where SG_{60} is the specific gravity (relative density) relative to water at 60°F (15.56°C) and 1 atmosphere pressure (101.325 kPa). This peculiar inverse definition was chosen because it gives a linear scale on a hydrometer. Confusingly, a higher density corresponds to a lower API gravity. Water has an API gravity of 10. The word gravity is unfortunate, because density actually has nothing to do with gravity, and there is a risk of confusion with gravitational acceleration. Most oil has an API gravity between 20 and 40. Heavy oil has an API gravity below 20, and extra-heavy oil an API gravity below 10. It must not be thought that API gravity is a complete description: two oils may have the same gravity but totally different compositions and other physical properties.

Figure 2.1 shows some of the first oil from well 34/7-P-29 of the Snorre field in the Norwegian sector of the North Sea. Figure 7.7 is a

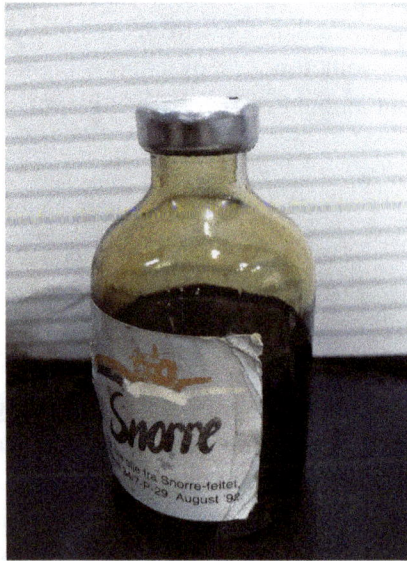

Figure 2.1   Crude oil from Snorre well 34/7-P-29

picture of the platform. Snorre oil is light (API gravity 37.5, defined above), low-sulphur (0.2% by mass), black in colour, and not very viscous. It is commingled with other slightly different crudes from the Statfjord area close by, and the resulting Statfjord Blend 2010 has the following properties (Oil and Gas Journal, 2015): the details are of little importance in the present context, but are included to show the range of data needed for a full description, and the unfamiliar ones are explained below

density at 15°C	0.8270 kg/l (827.0 kg/m^3)
specific gravity at 60°F (15.56°C)	0.8273
API gravity at 60°F (15.56°C)	39.5
Sulphur S	0.22% by mass
TAN (total acid number)	0.05 mg KOG/g
Reid vapour pressure	61.6 kPa
kinematic viscosity at 20°C	4.87 cSt (4.87 × 10^{-6} m^2/s)
nitrogen	890 mg/kg
hydrogen sulphide	not detectable
vanadium	1.1 mg/kg
nickel	1.1 mg/kg
sodium	7 mg/kg
wax	7.0% by mass
water content	0.06% by mass

Reid vapour pressure describes evaporation characteristics, and is the absolute pressure exerted by the oil at 37.8°C (100°F) in a standardised test. Total acid number is a measure of acidity, and is the mass of potassium hydroxide in mg needed to neutralise the acids in 1 g of oil: it is important because of corrosion. Kinematic viscosity is one of the measures of how viscous the oil is, and is the absolute viscosity divided by the density. The unit used to be centiStokes, a metric but not SI unit called after a nineteenth-century physicist, but the SI unit is m^2/s; 1 St (Stoke) is 10^{-4} m^2/s, and so 1 cSt is 10^{-6} m^2/s. The kinematic viscosity of water at 20°C is 1.002 × 10^{-6} m^2/s, and so Statfjord blend oil is about five times as viscous as water.

The numbers in the table show that the oil contains no hydrogen sulphide, so it is 'sweet' rather than 'sour'. The content of the heavy metals nickel and vanadium is extremely small.

Some crude oil is quite different. Heavy oil has an API gravity less than 22, although slightly different definitions are also used. Section 2.6 discusses heavy oil in more detail.

2.5 Hydrates

Hydrates are solid ice-like compounds of hydrocarbons and water (Sloan, 1998; Falser, 2012). Figure 2.2 shows a methane hydrate prepared in a laboratory: the flame is burning methane and radiation from the flame heats the hydrate and dissociates more methane. Figure 2.3 shows a gas hydrate formed in a gas pipeline. The gas molecules are held within a cage of water molecules. A hydrate is stable only at a combination of low temperature and high pressure. Figure 2.4 shows the stability boundaries for pure methane and for 90% methane and 10% ethane (an unusually high ethane fraction). A useful way of remembering roughly where the methane stability boundary is located is that at 4°C the pressure is 4 MPa.

Figure 2.2   Methane hydrate

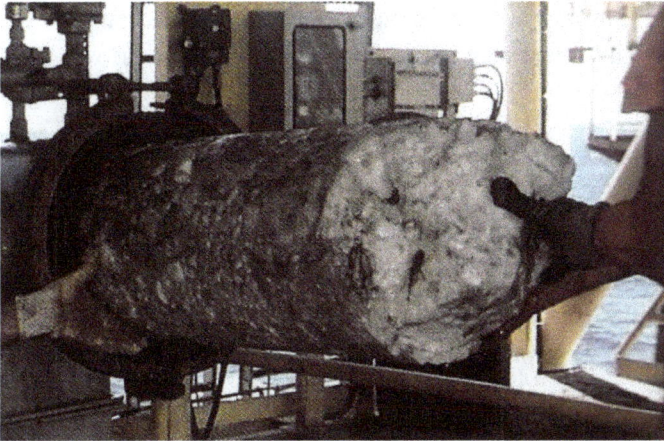

Figure 2.3   Natural gas hydrate

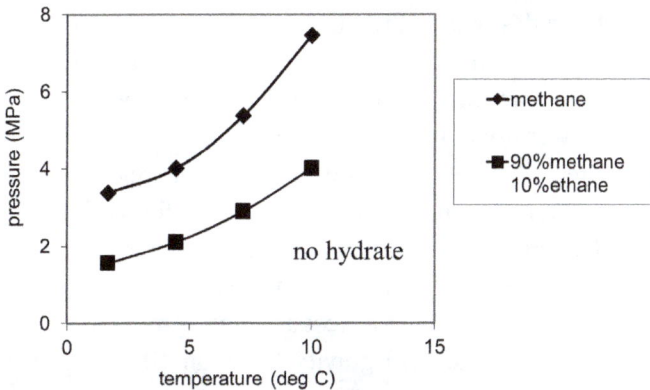

Figure 2.4   Stability boundaries for pure methane and 90% methane 10% ethane

Because to be stable they need low temperatures and moderately high pressures, gas hydrates are found in the Arctic and under deep water. There are very large hydrate fields in Canada, Siberia, Japan and many other areas.

Gas production from natural hydrates is discussed in Chapter 6. Natural hydrates will dissociate if global warming pushes the temperature past the stability boundary. The methane liberated is a powerful 'greenhouse gas', and that leads to a positive feedback that further increases warming. Hydrates are important in other contexts, because they can form within pipelines and process equipment and block the flow.

2.6 Heavy Oil

Heavy oil has an API gravity less than 20. Heavy oil is denser and more viscous than light oil, and has higher amounts of asphaltenes and resins, which are more complex hydrocarbons. The distinction is that asphaltenes dissolve in n-heptane and resins do not. Heavy oil often has a much larger percentage of sulphur, vanadium and nickel. Schlumberger (2006), Langevin *et al.* (2004) and Santos *et al.* (2014) are introductory reviews, and there are many others. Among the examples cited by Langevin, the gravities of Athabaska crude from Canada, Boscan from Venezuela, Cold Lake from Canada and Panucon from Mexico are 8.3, 10.2, 10.2 and 11.7° API respectively. Boduzynski *et al.* (1998) analyse the different components of heavy crudes, with particular emphasis on 13.6° API Kern River heavy oil from California.

Heavy oil is more difficult to produce, transport and refine than light oil, and has a lower price. It is much more viscous, and its chemical composition can create further difficulties: for example, the nickel and vanadium it contains can poison the catalysts used in refining. However, heavy oils are found in large quantities in many regions. Some oil companies and researchers have given a lot of attention to heavy oil, a valuable resource that will become more valuable as resources of lighter oils are used up.

Extra-heavy oil has an API gravity less than 10, so that it is denser than water, and a dynamic viscosity at reservoir temperature not higher than 10^4 cP; centipoise cP is a measure of dynamic viscosity. If the viscosity is higher still, the oil is classified as a natural bitumen. Extra-heavy oil is still more difficult to use efficiently. An Energy Council 2010 report lists the known reserves of extra-heavy oil. They are very large indeed.

Chapter 6 examines production.

2.7 Oil Sands

Oil sands used to be called tar sands, a name dropped because it is not strictly accurate and the word 'tar' provokes adverse reactions. They are sands that contain a mixture of sand, clay, water and a dense and viscous bitumen, and are found in huge quantities in northern Alberta in Canada, and in Russia and Kazakhstan. The sands in the Orinoco Belt of Venezuela are sometimes referred to as oil sands, but they are not bituminous and are more often thought of as heavy oil.

References

Alboudwarej, H., Felix, J. (J.), Taylor, S., Badry, R., Bremner, C., Brough, B., Skeates, C., Baker, A., Palmer, D., Pattison, K., Beshry, M., Krawchuk, P., Brown, G., Calvo, R., Triana, J.A.C., Hathcock, R., Koerner, K., Hughes, T., Kundu, D., de Cárdenas, J.L. and West, C., Highlighting heavy oil. *Oilfield Review*, **18**(2), 34–53 (2006) http://www.slb.com/~/media/Files/resources/oilfield_review/ors06/sum06/heavy_oil.pdf

Boduzynski, M.M., Rechsteiner, C.E., Shafizadeh, A.S.G. and Carlson, R.H.K. Composition and properties of heavy crude. UNITAR center for heavy crude and tar sands (1998).

Falser, S., Uchida, S., Palmer, A.C., Soga, K. and Tan, T.S. Increased gas production from hydrates by combining depressurization and heating of the wellbore. *Energy and Fuels*, **26**, 6259–6267 (2012).

Hunt, J.M. Petroleum geochemistry and geology. W.H. Freeman (1996).

Santos, R.G., Loh, W., Bannwart, A.C. and Trevisan, O.V. An overview of heavy oil properties and its recovery and transportation methods. *Brazilian Journal of Chemical Engineering*, **31**(3), 571–590 (2014) http://www.scielo.br/scielo.php?script=sci_arttext&pid=S0104-66322014000300001.

Sloan, E.D. and Koh, C. Clathrate hydrates of natural gases. CRC Press (1998) World Energy Council. 2010 Survey of energy resources. www.worldenergy.org (2010).

Chapter Three

Where Does Petroleum Come From, and Where is it Now?

3.1 Introduction

This chapter examines how petroleum was formed, and how it arrived at where it is now, not usually the same place. It begins with an elementary introduction to the geophysical background, and goes on to consider the generally accepted theory of hydrocarbon formation, held by the great majority of geologists. Petroleum is formed from biological material by the effect of millions of years of high temperature and high pressure. It is formed in a rock we call a 'source' rock, but if we are to find and exploit it is usually has to migrate to another permeable rock, called a 'reservoir' rock. If it is to be useful, though, something has to stop it migrating all the way to the surface, because then it will form a seep and escape into the atmosphere or the sea, and that indicates the categories of geological conditions that are like to correspond to a useful and attractive hydrocarbon reservoir. There is an alternative theory of formation, held by a small minority of scientists, and that is briefly set out in section 3.7.

3.2 Background in Geology

The centre of the Earth is much hotter than the surface. The difference is explained by continuous heat generation by radioactive decay, and to a lesser extent by remnant heat from the Earth's original formation as a mass of molten rock. Heat is conducted outwards from the centre to the surface. As you go down from the surface, the temperature increases at about 0.03°C/m, but more rapidly in volcanic regions. The difference is large enough so that you can sometimes feel it if you go down a tunnel or into a mineshaft.

The pressure also increases with depth, because at any level the pressure has to carry the weight of the rock or soil above it. If the rock is a 'solid' continuous single phase like granite, the weight has to be carried by the vertical components of the stress. Those components vary horizontally and vertically, over both small distances (0.1 m) and large (10 km), but together they keep in the overlying rock in equilibrium.

If the soil has two phases, like a sand or a sandstone made up of solid particles of sand minerals with the pore spaces between the particles filled with water or another fluid, then the load has somehow to be shared between the stress carried by the particles and the stress carried by the fluid. Geotechnics calls the first component the effective stress and the second component the pore pressure. Together the two components add up to the total stress. It is that total stress that has to maintain equilibrium. If the soil is distorted, both effective stress and the pore pressure change. If the pore fluid can move through the soil, then the pore pressure can change further: if for example a distortion creates a high pore pressure, that pressure can leak away into another region with a lower pore pressure. The effective stress concept is discussed and illustrated at much more length in standard texts on geotechnics: see, for example, Terzaghi and Peck (1948) and Schofield and Wroth (1968). The idea can be extended to soils with more than one fluid phase, for example water and natural gas together.

Geology makes a distinction between three kinds of rock, igneous, sedimentary and metamorphic. Igneous rocks (from the Latin 'ignis' for 'fire') are formed by the cooling and solidification of molten magma and lava. Chemical and physical processes break up igneous rocks, and water and air move the broken fragments and deposit them somewhere else. The fragments may then coalesce to form a 'solid' sedimentary rock. Examples are sand dunes, where the particles have been moved by air or water, remain loose and have not connected at all, or alternatively sandstone, where continuing chemical and physical alteration has formed a continuous porous material that may have substantial strength. A sedimentary rock may also be formed by deposition of solids from solutions: for example, calcium carbonate can precipitate from a solution that contains dissolved calcium molecules and carbon dioxide, and the carbonate forms a solid limestone. A sea may dry up and deposit salt as a solid. Metamorphic rocks originate from sedimentary rocks by continuing chemical physical processes, which lead to internal recrystallisation and dramatic changes in properties. Marble is metamorphosed limestone, and often incorporates coloured veins that originate from other minerals such as iron which migrated into the limestone. Anthracite is metamorphosed coal. Slate is a metamorphosed shale, and often the clay minerals have recrystallized under strong compression to align and form the planes of weakness that allow slate to split easily.

All these processes occur over tens of millions of years. They are often repeated. A rock may solidify as an igneous rock, then be broken up and deposited as a sediment, then be heated to become a metamorphic rock, then eroded and deposited again as a different sediment, and so on. It is this that accounts for the richness of geological possibility.

Over time, the Earth distorts. Some of the movements are on a very large scale. The continents move relative to other, over thousands of kilometres, and that movement is continuing today. North America is moving away from Europe at about 25 mm/year, the speed your fingernails are growing at. Instructively, the theory of that continental drift was thought by almost all right-thinking geologists to be complete nonsense when it was first put forward by Wegener a century ago. He had noticed that the coastlines of Africa and South America would fit together almost perfectly, as you can see by looking at a globe, and later investigation showed that the fit is still better if the coastlines are taken from contours at a different water depth. More evidence accumulated. The theory is now accepted by everyone.

Some of the movements are on a smaller scale. The Earth can distort smoothly, as it does if you squash a lump of butter or Blu-Tack between your fingers, or it can break, as it does if you try to bend a brittle cookie. Both kinds of movement can occur in the same material. Ice deforms smoothly if it is loaded by small stress, as it is in a glacier, or it can fracture, as it does if you hit an ice-cube with a hammer or try to skate on thin ice. The breaks are called 'faults', and can go in different directions relative to the stresses that caused them. Geology has an elaborate vocabulary to describe different kinds of faults.

3.3 Formation of Petroleum

Petroleum is formed by the action of heat and pressure on sedimentary rocks that contain biological materials from animal and plant life. The places where we find petroleum now are not generally the places where it was formed.

Animals and plants make up the web of life, and exist over an immense range of sizes, from elephants and large trees down to bacteria and viruses. When they die, their remains are carried into the sea or rivers or lakes, and fall to the bottom, though on the way they may form part of the food chain of other kinds of organisms. Sediment falls on top of the remains of life, and they are progressively buried, more and more deeply as more sediment arrives, together with other remnants that arrive later. As the animal and plant remains are buried in the sediment and get to be deeper below the surface, the temperature and pressure increase, by the processes outlined in section 3.2.

At small distances below the surface, the temperature is only a little above the surface temperature. Bacterial action begins, and the organic material decomposes. Methane gas is released, as it is when we stir the bottom of a stagnant pond, or in our own digestive systems. That gas is called biochemical methane. Further down, the temperature is higher, and the organic material becomes kerogen. Kerogen is not a specific compound, but the word is used as a general terms to describe various compositions. Hobson and Tiratsoo (1981) describe kerogen as 'a dark-coloured coaly material, insoluble in acids and organic solvents, composed of bacterially-altered organic detritus with algal and bacterial remains'. Type 1 kerogens are derived from algae, have large amounts of fats ('lipids'), have high hydrogen and low carbon, and derive from algae. Type 2 kerogen has less hydrogen and more oxygen compared to type 1, and derives from plankton. Most oil and gas derives from type 1 and type 2. Type 3 again has less hydrogen and more oxygen, and derives from plant cells (Hobson and Tiratsoo (1981)).

At higher temperatures, kerogens progressively transform. At about 60°C, types 1 and 2 split and release smaller hydrocarbon molecules, the methane, higher alkanes such as ethane and propane ('wet gas') and liquid petroleum. The temperature range from 60°C to 160°C is often called the 'oil window'. Beyond the top of that range, the released

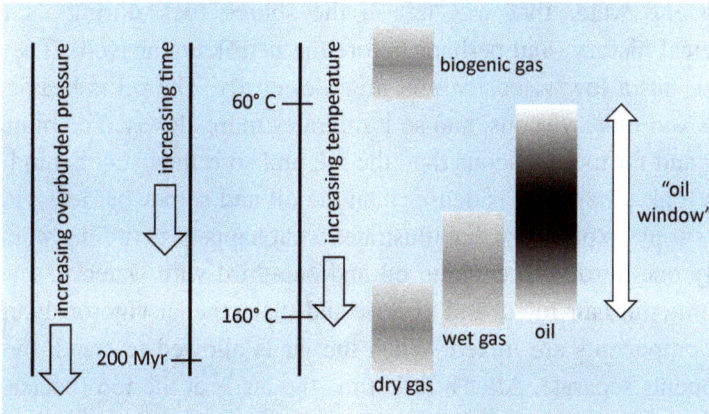

Figure 3.1   Schematic history of petroleum formation

hydrocarbons are principally methane, and further on any remaining hydrocarbons split into methane and carbon. The process is illustrated schematically in Figure 3.1.

More elaborate forms of this diagram appear in texts: see for example, Hunt (1996) and Hobson and Tiratsoo (1981), and they name the different stages and relate the progressive transformations to changes in oxygen/carbon and hydrogen/carbon ratios.

It needs always to be remembered that these processes occur over tens of millions of years. They can only partially be simulated a laboratory, where the timescale is 10^7 times shorter.

3.4 Migration

The petroleum usually migrates from the 'source rock' in which it was formed to the 'reservoir' where we find it today.

Migration may be both vertical and horizontal. The driving forces for migration are gravity and the pore pressure changes induced by tectonic movements, by temperature changes, and by changes in total stress created by the increasing depth of sediment above the source rock. The different components of the petroleum may migrate at different speeds and in different directions. Consider for example a petroleum that has three components simplified to gas, oil and bitumen, accompanied by a fourth component, water that was left in the source rock during its earlier geological history (and perhaps before the petroleum arrived). The gas is lightest, has a low viscosity, and migrates easily. The oil is heavier than the gas and more viscous, and so it migrates more slowly. The bitumen is denser and far more viscous than the oil, and so it may stay behind in the source rock. The water is denser than the oil and somewhat less viscous.

A simple experiment that illustrates what happens is to fill a storage jar roughly one-third with cooking oil and one-third with water, to leave the remaining third air-filled, and to close and shake the jar vigorously until all three components are mixed. When the jar is allowed to stand, the three components separate. After a little time the air is at the top (because it is least dense), the oil is in the middle (because it is denser than the air but less dense than the water), and the water is at the bottom. If instead we repeat the experiment but before starting it fill the jar with fine sand, then the final separation between the three fluid components will be nearly the same, but it will take much longer. Some bubbles of gas will remain in the lower oil and water components, because the bubbles are held back by surface tension and the density difference is not large enough to drive them up. Similarly, some drops of oil will remain in the water, and some drops of water will remain the oil. That too is what happens in a reservoir.

Figure 3.2 is a heavily-idealised schematic of a reservoir that contains oil, gas and water. If there is no gas or no oil, those components will of

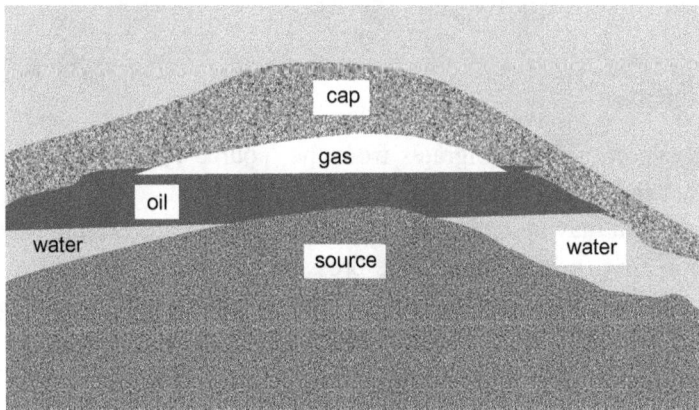

Figure 3.2   Schematic of a simple reservoir

course be missing. A sketch of that kind oversimplifies the boundaries between the phases: close to the boundary between oil and water, there will be some water in the oil and some oil in the water.

Surface tension plays a part. In a fine-grained material, the forces induced by surface tension may be large enough to stop movement of the fluids, even though a density difference might otherwise drive flow. The relevant surface tension may be between a fluid and a solid particle, or between one fluid and another.

Figure 3.3 is a photomicrograph of Ravelstone Black Sandstone from a quarry near Corstorphine in Scotland. The pores are partially filled with black petroleum. This is not the source rock, and it is thought that the petroleum had migrated from elsewhere.

3.5 Traps

The formation and migration processes described in sections 3.3 and 3.4 are not by themselves enough to create a reservoir that we can exploit. If there is nothing to hold the petroleum fluids where they are, they can go

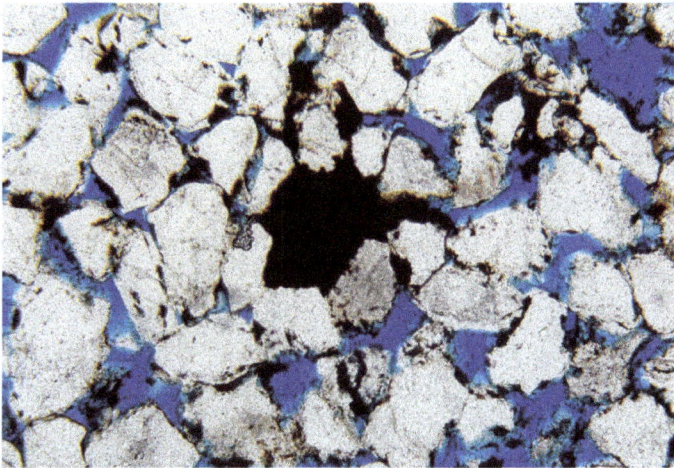

Figure 3.3   Photomicrograph of Revelstone Black Sandstone
Courtesy of the British Geological Survey, Catalogue Number: P526984

on moving, and in the end they will reach the ground surface and be dispersed into the air or water. To create a reservoir we need a third component in addition to the source and the reservoir. The third component is a 'cap' that stops the petroleum moving further. Together they form a 'trap'.

Figure 3.2 is a schematic that illustrates one kind of trap. The gas has risen until the cap stopped it rising further. The oil is below the gas because it is denser than the gas, and the water is below the oil because it is denser than the oil. A sketch of that kind oversimplifies the boundaries between the phases, because it misleadingly suggests that the boundaries are sharp. In reality there is a transition between the phases, so that close to the oil/water boundary there is some oil within the water, dissolved or in droplets, and some water within the oil.

Sketches like Figure 3.2 are necessarily two-dimensional, but the world is of course three-dimensional. If a cap is to be effective, it has to enclose the petroleum all round, not just at the sides as we see them in two dimensions.

A cap has to be comparatively impermeable. A typical cap rock is 10^4 times less permeable than the reservoir below it.

Traps exist in almost infinite variety. Figures 3.2 and 3.4 through 3.8 illustrate four possibilities.

The first is an anticline, the geological term for a 'hill' in the strata, with a reservoir of both gas and oil that occupies the top of the anticline, shown in Figure 3.2. Figure 3.4 is a photograph of an exposed section of an anticline, at Beinn Eighe near Loch Maree in Scotland (but not in an oilfield). This and other photos were chosen because the geology is exposed and simple and shows up clearly.

The second is a trap created by a fault. Figure 3.5(a) on the left shows an unfaulted sedimentary rock, and on the right (b) the same structure after a fault has run through it. The sediments on the right have moved upwards relative to the sediments on the left. That has created a seal against one of the finer sediments within the formation to the right, and oil has risen and been trapped. Figure 3.6 is a photograph of an exposed fault at Coire nam Beithach near Glencoe in Scotland. A continuation of the same fault can be seen in the distance, a few km away half-way down the right slope of Bidean nam Bian.

Figure 3.4   Exposed section of anticline

Courtesy of the British Geological Survey, Catalogue Number: P001674

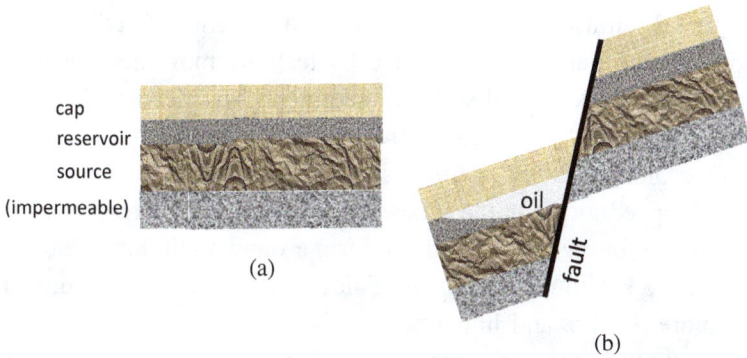

cap
reservoir
source
(impermeable)

oil

fault

(a)

(b)

Figure 3.5 Faults (a) before faulting (b) after faulting

A fault is not necessarily a seal. The gap between the 'solid' sediments on either side may be filled with a coarse sediment that has a relatively high permeability, and in that case the fault will not create a reservoir and instead form a pathway for petroleum to move higher.

Figure 3.6   Fault Courtesy of the British Geological Survey, Catalogue Number: P000260

Figure 3.7 illustrates a third kind of trap. A series of sediments were laid down, and then they were tilted by tectonic movements and partly eroded by water. A second series of sediments was deposited, on top of the first but at a different angle, creating an unconformity. If petroleum can move along coarser sediments in the first series, it can come up against finer sediments in the second series, and forms a reservoir.

Finally, a fourth kind of trap can form around a salt dome. Salt is laid down when a shallow sea dries up. Salt is lighter than most sedimentary rocks, more ductile, and impermeable. Over geological time, a mass of salt rises through the sediments above, forming a dome. As it rises, it pushes up the sediments above it, forming the traps seen in Figure 3.8. Similar structures can be created by volcanic intrusions.

A classic example of an oilfield centred on a salt dome is Spindletop, in Texas north-east of Houston. The surface expression of the dome is quite modest, a hill no more than 10 m high. The area was known for gas seeps and sulphurous springs, and it had been suspected that oil might be

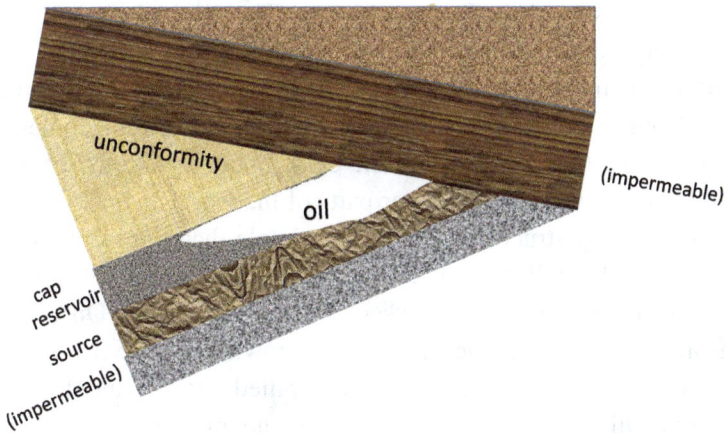

Figure 3.7

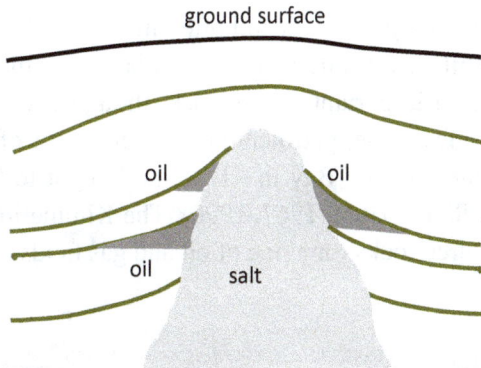

Figure 3.8

there. Several wells had been drilled, either completely without success or producing only tiny quantities. Another well was drilled and broke into the reservoir on January 10 1901, at a depth of 347 m. A spectacular gusher flowed 100,000 b/d until it was brought under control nine days later. At Spindletop the cap is gypsum and anhydrite, and solution by water forms cavities, so that when a cavity fills with oil under pressure and is drilled into, the oil flows rapidly.

Spindletop started the Texas oil boom. The original part of the field declined very rapidly because of overproduction from hundreds of wells, but further drilling found larger reserves in other parts of the structure. Many other salt domes were drilled soon afterwards. Salt domes occur elsewhere along the Gulf Coast of the USA, under the North Sea, in Iran, Iraq and Arabia, and in many other parts of the world.

Most real trap structures are far less simple than the idealised examples illustrated here. Every one of them is different. Often different kinds of traps occur together or are stacked one above the other. Oil or gas can spill from one trap to another, in the way shown in Figure 3.9, where the left-hand anticline is full and the oil has spilled into the anticline further to the right. Oil can also leak upwards from one imperfect trap to the next.

3.6 Shale Oil and Gas

Shale is composed of extremely fine particles of quartz and clay minerals like kaolin, less than 4 μm across and often much smaller. That gives it a very low permeability. Organic material was buried as the minerals sedimented. Shale is an important source rock, though it can also be a cap rock. Over geological time, petroleum can migrate out of the shale. The huge oil field near Prudhoe Bay in Alaska is thought to have originated from the Kingak Shale (Hunt, 1967, 1996). The Kimmeridge Shale is one of the primary source rocks for a line of oil and gas fields that stretch from

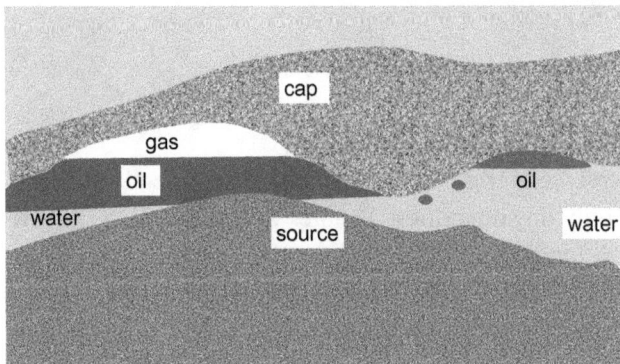

Figure 3.9 Movement between traps

Wytch Farm in the south of England to the Norwegian sector of the North Sea. The Bazhenov Shale is the source of the huge Salym field in Siberia.

Shale can be mined and burned, but only if it is practicable to excavate it from shallow formations. Shale was first reported ten centuries ago, and has been mined for many years. Burning creates large quantities of ash and therefore environmental problems. Estonia is the only country to derive most of its energy from mined shale, and there it accounts for 85% of national electricity production. China has large reserves.

It used to be thought impracticable to produce oil and gas from shale by conventional techniques, because the permeability is so small. Within the last ten years, that opinion has dramatically changed. The technical background is that it has been found that wells can be drilled horizontally rather than vertically, and that the shale around a horizontal well can be extensively fractured ("fracking") by applying high pressure and holding the fractures open. The fractures themselves are highly permeable, and their presence shortens the drainage paths through which the petroleum can reach the well. The technology is discussed further in Chapters 5 and 6.

The USA has taken the lead, and as a result has become a net exporter, with revolutionary economic and political impacts. A combination of oversupply, economic and political factors has caused the oil price to drop precipitously, from more than 100 $/barrel to 27 $/barrel at the time of writing. That drop has had a severely damaging short-term effect on the shale oil and gas boom, but it can be expected to recover as the oil price increases.

3.7 Abiogenic Theory of Petroleum Formation

There are alternative theories, but they are not widely accepted. 'Abiogenic' means without the engagement of life.

The iconoclastic physicists Thomas Gold (1920–2004), Fred Hoyle (1915–2001) and some Russian scientists have argued that the accepted biological theory of the origin of natural gas is incomplete, and that more gas is to be found much deeper into the Earth's crust. Hoyle and Gold were extremely distinguished in other fields, among them cosmology, astronomy, radar and hearing. Their opinions cannot be rejected out of hand. Hoyle was notoriously rude, and later in his career became cranky. He is said to have remarked sarcastically that

> 'The suggestion that petroleum might have arisen from some transformation of squashed fish and biological detritus is surely the silliest notion to have been entertained by substantial numbers of persons over an extended period of time.'

In Gold's famous words,

> 'Hydrocarbons are not biology reworked by geology (as the traditional view would hold) but rather geology reworked by biology'

Gold argued that since the hydrocarbons that form the major component of petroleum are present across the solar system and elsewhere in the universe, there is no need to believe that on the Earth they must have a biological origin. He proposed that the molecules that petroleum is formed from were trapped about the time when Earth was formed, several billion years ago, and that they encountered microbes as they moved upwards towards the surface. An attempt to test that hypothesis was made by drilling at Lake Siljan in Sweden into a meteor crater that would have opened up pathway channels down to 40 km. The Gravberg 1 well brought up 0.1 m^3 of oily sludge. A second well to 6.8 km depth recovered 13 m^3 of oil, not remotely enough to be commercial. Sceptics argue that the oil was contamination from drilling.

Most of the petroleum geochemistry community remains unconvinced or hostile. Hobson and Tiratsoo (1981) discuss Gold's theory and other abiogenic theories, but Hunt's magisterial book on petroleum geochemistry (Hunt, 1996) does not bother to mention Gold or any abiogenic alternative. Freeman Dyson wrote that

> 'Gold's theories are always original, always important, usually controversial — and usually right'

but he did not say that they were always right. The reader must make up her own mind about whether this kind of theory is worth pursuing.

References

Hobson, G.D. and Tiratsoo, E.N. Introduction to petroleum geology. Scientific Press (1981).

Hunt, J.M. Petroleum geochemistry and geology. W.H. Freeman (1996).

Schofield, A.N. and Wroth, C.P. Critical state soil mechanics. McGraw-Hill (1968).

Terzaghi, K. and Peck, R.B. Soil mechanics in engineering practice. Wiley (1948).

Chapter Four

Finding Petroleum

4.1 Introduction

Petroleum has to be found before humanity can use it. A petroleum field gets used up, sometimes in a few years and sometimes over many decades, and so we have to keep finding more. This is another huge subject, and hundreds of thousands of people are engaged in it. The potential rewards are so great that much research is carried out, and the technology changes.

Survey techniques make use of different physical and chemical techniques to locate and delineate petroleum reservoirs. Almost every area of physics has been experimented with. Some techniques are 'active': they send some disturbance into the ground, and observe the response. Seismic survey is active, and much the most important: it sends a stress wave into the earth, and measures the reflected waves that come back. Other techniques are 'passive': they measure different kinds of signals naturally radiated from beneath the surface: magnetic, gravity and geochemical surveys are examples.

4.2 Seeps and Anticlines

Seeps have been known about for thousands of years. Oil seeps are visible, and humanity soon discovered that the seeping fluid can be burned. The oil seeps in ancient Iraq are mentioned in the Bible. In China 2000 years ago, the historian Ban Gu wrote that

> "The waters in the You River in Gaonu (Shanxi) are rich enough to burn. The fat on the water can be collected for direct use'

and in the eleventh century Shen Kuo (1031–1095) wrote

"…There is petroleum (stone oil) in the Yan territory, generated among sand and stones in water. It burns like flax, but produces dense smoke… this substance will be of great use and popularity in the future…" (Zigong Salt Industry Museum, no date; Sun, 2005)

Stone oil is still the Chinese word for petroleum. Similar stories come from many places. The Dutch planter Zijker recognised oil on ponds in Sumatra in 1880, saw that the local people were using the oil, and went on to develop one of the precursors of Shell. Later the explorers Brower and Leffingwell (1875–1971) found oil seeps on the Arctic coast of Alaska, and the Spindletop field in Texas [chapter 3] was identified because of seeps.

Seeps can be quite small, but they may still indicate the presence of large amounts in the past. On the beach at Lulworth Cove in southern England there is a small oil seep just above high-water level. If a hundred people walk past, ninety-seven do not notice it, two think that some fool has dumped the old oil from his car, and the last one wonders if it might be natural. Guessing that the flow is 0.0001 m^3/day, and that it has continued for 150 million years (another guess), the volume of oil released over that time was 36 million barrels, enough to be interesting. It may of course almost all have gone. Lulworth has not been drilled, but it is close to Wytch Farm, the largest onshore oilfield in Western Europe.

There are gas seeps, but they are not easy to see. There are both oil and gas seeps under water. Gas is collected off Santa Barbara in California by placing two steel pyramid-shaped tents over a group of gas seeps.

Hunt (1996) cautions against expecting too much from seeps or from the near-surface presence of oil:

"Oil and gas seeps occur throughout the Western Canada Basin. However, it took 400 dry holes before the Leduc discovery made Canada a major oil producer…The history of oil exploration is replete with examples of dry holes being drilled in the vicinity of seeps until luck or ingenuity made the big discovery. Neither visible seeps nor surface prospecting can unequivocally outline a petroleum accumulation at depth, except where the accumulation is shallow and its caprock is leaking directly upward…"

People who found petroleum noticed that it was often found in anti-clines [3.5]. An anticline with seeps was therefore a promising target, if the topography suggested that any reservoir would be 'closed', so that the petroleum could not migrate out in any direction. Most of the world's oil is found anticlines, but it is impracticable to drill every anticline. Moreover, by no means every anticline contains petroleum, and an anti-cline far beneath the surface may not create a recognisable feature at the surface.

An exploration geologist would therefore like to find a way of looking below the surface without drilling. He can use several physical phenom-ena. Often they can be combined, just as we use different senses to register the world around us. If we are frying eggs, and hear a loud sizzling, smell butter, and see smoke, then we integrate the information from hearing, smell and sight, and recognise that the eggs are too hot and are burning. Our existing knowledge adds to that interpretation: we know that we are frying eggs rather than sausages, and we think it unlikely — though not impossible — that the sensory inputs are arriving because the building is on fire.

4.3 Gravity Surveys

Any two masses have a gravitational attraction. The attractive force is pro-portional to the product of the two masses divided by the square of the distance between them. The gravitational force we feel at the surface is the integrated effect of all the mass within the Earth. Except for that, we are not aware of gravitational attraction generally: there is a gravitation attrac-tion between two books on a desk, but it is far too small to measure.

The gravitation acceleration g (the acceleration of a mass falling freely) is not quite uniform across the surface. One reason is that the Earth is not perfectly spherical, but is slightly flattened at the poles, so that g at sealevel is 9.78 m/s^2 at the Equator but at the North Pole 9.83 m/s^2. There are also local effects. Consider for example a salt dome [3.5, Figure 3.8], created by salt slowing rising through the sediments above it. The salt is lighter than the sediment, which is why it rises, and so a mass above the dome will be slightly less attracted, because the mass of the salt is smaller than the mass of the sediment it displaced. That will reduce the

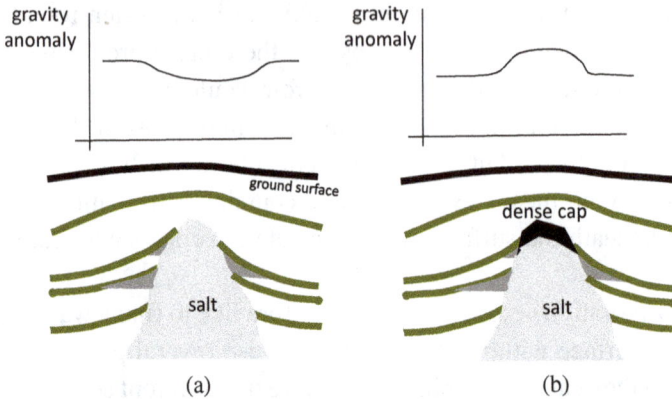

Figure 4.1   Gravity anomalies above salt domes

gravitational acceleration above the dome, in the way shown in Figure 4.1(a). However, a complication is that if there is a dense cap rock above the salt, the effect of that cap is to increase the gravitational acceleration above the dome, as shown in Figure 4.1(b). An igneous intrusion can create the same effect. The differences are very small, typically 0.00001 m/s^2.

Gravitational measurements have been applied successfully to identify salt domes and other features. They can be used from aircraft and ships, and are widely applied to minerals exploration. Nabighian *et al.* (2005) is a useful review of the whole subject.

4.4 Seismic Surveys

Seismic surveys are called after the Greek for 'earthquake'. An explosion or dropped weight at the surface creates stress waves. The stress waves travel into the soil and rock below, and are reflected when the elastic properties alter, at a shift from one rock type to another. Measurements and interpretation create a picture of the structure.

Stress waves in solids are more complicated than sound waves in a fluid. A solid can transmit shear waves as well as pressure waves, and in addition there are other waves at a surface (Kolsky, 1953; Johnson, 1972), but seismic survey principally depend on waves of pressure.

Several books take the subject much further: see, for example, Chapman (2004), Ikelle (2005) and Vermeer (2012).

Imagine a level ground surface with a level layer of rock 1 immediately below the surface, and a level layer of a different rock 2 at some depth (Figure 4.2 (a)). A sharp explosion at point A generates a stress wave. The wave travels downward through rock 1 and strikes rock 2. At the interface there is a mismatch of elastic properties. Some of the seismic energy is reflected back towards the surface, and some continues downward into rock 2 (at a slightly different angle because of refraction). When the reflected wave arrives back at the surface, there is another mismatch between rock 1 and the air, another reflection down to the interface, another reflection from the interface, and so on. The reflections progressively weaken, because energy is absorbed within the soil and because of radiation in other directions. In the Figure, the strengths of the different waves are indicated by the breadths of the lines.

The movements are measured by a geophone, which is like a microphone but set up to respond to movements of the ground rather than the air. The version used at sea is called a hydrophone. Figure 4.2(b) plots the time history of the movements recorded by a geophone at point B, close to point A. The time from the initial explosion to the arrival of the first reflection is called the two-way travel time. If the velocity of sound in the soil is known, the depth to the interface is half the two-way travel time multiplied by the velocity.

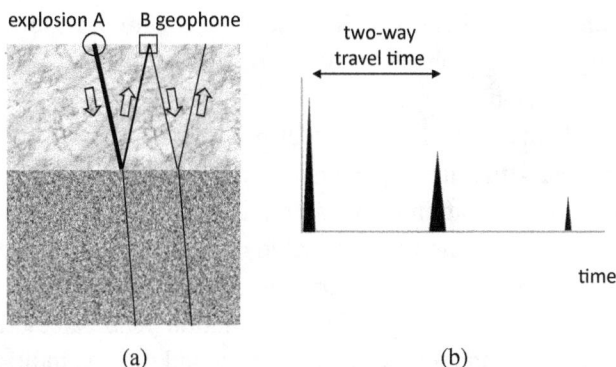

Figure 4.2   Reflections from an interface to one geophone. (a) wave paths (b) time record at geophone B

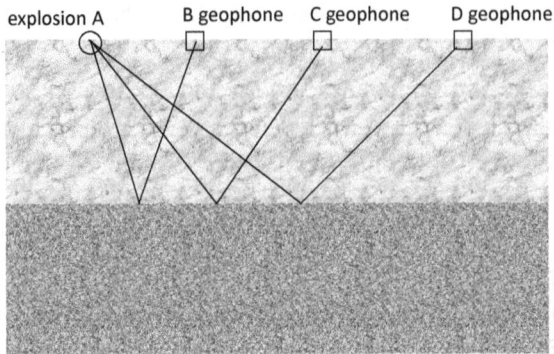

Figure 4.3   Reflections from an interface to three geophones, above a horizontal interface between rock 1 and rock 2

Imagine instead that there are three geophones B C and D along a straight line from the explosion point A (Figure 4.3). The distances between them are known. The two-way time to C will be larger than the two-way time to B, because the distance is longer. If we can assume that the interface is horizontal and the soil is uniform, and we apply a little simple geometry, the two two-way times will give us both the depth to the interface and the wave velocity in rock 1. The two-way time to D will give us a check.

Next, imagine that the interface is not horizontal but sloping (Figure 4.4). The distances are different again, but using the three two-way times we can calculate the velocity, the depth and the slope.

Those are idealised and simple structures. Figure 4.5 shows a simple unconformity, three geophones, and multiple wave paths from the explosion point and various reflector interfaces to geophone C.

The complexity of the record will clearly increase with the complexity of the subsurface structure and the number of geophones. One factor that helps with the interpretation is that all the geophones must be 'seeing' the same structure. Geophone records include an enormous amount of data. Their interpretation into a usable form used to be done visually (Hobson, 1981), but is now accomplished by computation. The calculation eliminates various distorting factors, such as horizontal offsets, multiple reflections, surface waves and noise. Figure 4.6 is an example of a structure arrived at in this way.

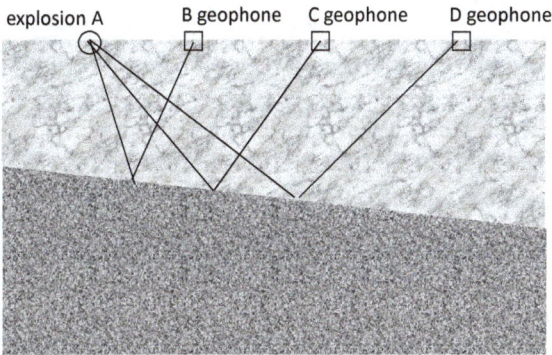

Figure 4.4   Reflections from an interface to three geophones, above a sloping interface between rock 1 and rock 2

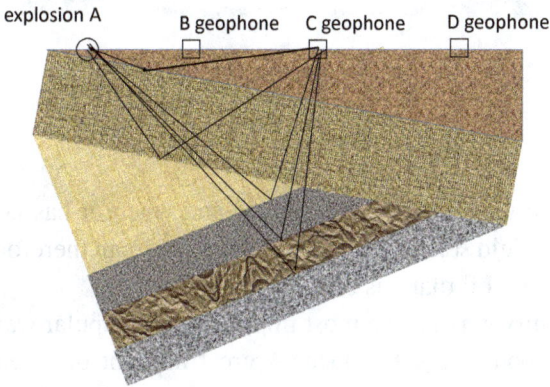

Figure 4.5   Reflections from several interfaces to three geophones

The simplified analysis so far has looked at two-dimensional models. The world is obviously three-dimensional, and the analysis can be extended by setting out a grid of geophones ('3-D seismic'). In seismic surveys at sea, that is done by having a survey vessel tow several 'streamers', lines several km long that carry hydrophones at regular intervals.

A further development looks at the evolution of a reservoir as a result of production. When oil is produced, some of it is bypassed and left behind, so that in some sections of the reservoir the oil replaced by water and in others the oil remains in place. That change alters the seismic properties. If therefore two surveys are made a few years apart, it becomes

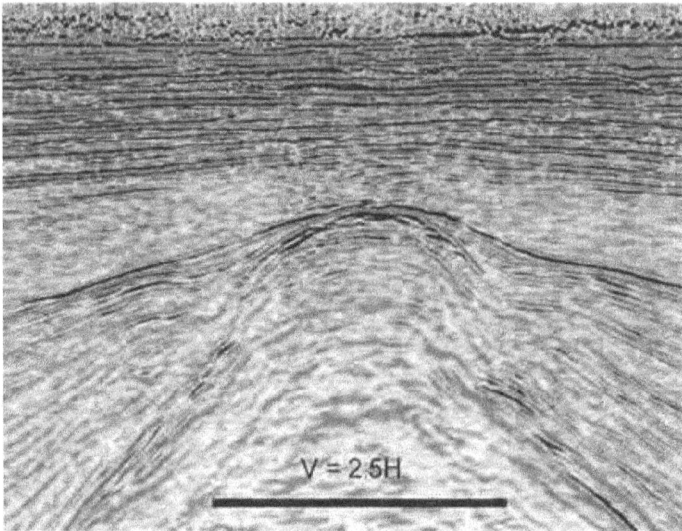

Figure 4.6

possible to see where the oil has moved and where it has not ('4-D seismic', 'Life of Field seismic'). A drilling program can therefore be planned so as to intercept oil that has been bypassed.

Seismic survey is far the most important and popular way of securing information about the subsurface. A great amount of research effort is given to it. It is important in other fields such as mineral exploration and submarine detection. It is not without limitations: in particular the spatial resolution is limited.

4.5 Magnetic Surveys

The Earth is a magnet, and everywhere there is a magnetic field. The direction of the field is different at different points on the surface, and slowly changes with time. A magnetic compass needle aligns itself with the horizontal component of the field. Magnetic rock such as ilmenite and magnetite disturbs the natural field, and alters its direction and amplitude. Sensitive magnetometers can measure those changes, and interpretation can build up a picture of structures subsurface. The method can be used from aircraft. The changes are again very small, and have to be corrected for the magnetic field generated by the aircraft engines and by steel structures.

4.6 Surface Geochemistry

A little petroleum can sometimes migrate from a reservoir to the surface, even if there is no visible seep. A possible option is to take samples over an area, and analyse them for hydrocarbons. Hunt (1996) explains some of the difficulties [4.2], and refers to Philp (1987) for a detailed review. In particular, hydrocarbons migrate from source rocks as well as reservoirs, and do not necessarily migrate vertically.

4.7 Drilling

If the results of the surveys [4.2–4.6] show that there are structures that might contain worthwhile amounts of petroleum, and if the estimates of volume indicate that production might be economically attractive, then somebody can decide to drill. Drilling is much more expensive than survey, and success is far from certain. Many holes are 'dry'. In 1982 BP drilled the Mukluk well from an artificial gravel island off the Arctic coast of Alaska. The well cost more than 1 B\$, and was dry. The oil that was once in a reservoir under Mukluk is now thought to have migrated to the Kuparuk field.

Drilling makes it possible to secure more geological and geochemical information. The vertical and horizontal resolution is much better, down to a few cm. Cores can be chemically analysed, and microfossils in the cores allow the formations to be dated. Hydrocarbons can move from the formation into the drilling mud [5.5], and can be detected by mud logging.

Drilling technology is described in Chapter 5.

References

Chapman, C.H. Fundamentals of seismic wave propagation. Cambridge University Press. (2004).

Hobson, G.D. and Tiratsoo, E.N. Introduction to petroleum geology. Scientific Press 3D.

Hunt, J.M. Petroleum geochemistry and geology. W.H. Freeman (1996).

Ikelle, L.T. and Amundsen, L. Introduction to petroleum seismology. Society of Exploration Geophysicists. (2005).

Johnson, W. Impact strength of materials. Edward Arnold (1972).

Kaye, G.W.C. and Laby, T.H. Tables of physical and chemical constants. Longman (1990).

Kolsky, H. Stress waves in solids. Clarendon Press, Oxford (1953).

Nabighian, M.N., Ander, M.E., Grauch, V.J.S., Hansen, R.O., LaFehr, T.R., Li, Y., Pearson, W.C., Peirce, J.W., Phillips, J.D and Ruder, M.E. The historical development of the gravity method in exploration. *Geophysics*, **70.6**, 33–61 (2005); utam.gg.utah.edu/edhelper/papers/grav.pdf

Pearson, W.C., Peirce, J.W. Phillips, J.D. and Ruder, M.E. The historical development of the gravity method in exploration. http://utam.gg.utah.edu/edhelper/papers/grav.pdf. (2015)

Philp, R.P. Surface prospecting methods for hydrocarbon accumulations. *Advances in Petroleum Geochemistry*, **2**, 210–253 (1987).

Sun, J.S. A city with salt wells everywhere. Guangxi Normal University Press (2005).

Vermeer, G.J.O. 3D seismic survey design. Society of Exploration Geophysicists, Tulsa (2012).

Zheng, J. and Palmer, A.C. Bamboo pipelines in ancient China — and now? *Journal of Pipeline Technology*, **8** 95–98 (2009).

Zigong Salt Industry Museum (in collaboration with Shell). Drilling and gas recovery technology in ancient China. Zigong Salt Industry Museum, Zigong, Sichuan, China (2005).

Chapter Five

Drilling

5.1 Introduction

This chapter outlines the techniques applied to drill for petroleum. Drilling is an expensive and risky component of petroleum production, and has been given a lot of attention. Nguyen (1996) is an excellent book on the subject.

Cable drilling is considered first. In China it has been in use for thousands of years, and much later it drilled the famous Drake well that started off the petroleum industry in the USA. It is nowadays almost obsolete in the petroleum context, and has been replaced by rotary drilling. Figure 5.1 illustrates the two methods. A complete picture of a drilling rig is complex and confusing, and so the approach taken below is to examine the different components separately, and then to integrate them.

5.2 Cable Drilling

Cable drilling ('cable-tool drilling' 'percussion drilling') was developed in China some 2000 years ago. It drilled for salt and the natural gas found with it. The gas was burned to evaporate the salt brine. The drilling principle is illustrated in Figure 5.1(a). A heavy iron chisel is repeatedly lifted and dropped, and the impact breaks the rock at the bottom of the hole. In a separate operation, the broken fragments are removed by lowering a bailer, a tube with a one-way valve at the bottom. The fragments flow into the tube, carried by the water that has partly filled the hole, and are lifted to the surface.

In the original version of the system, the chisel was supported by a cable attached to a cantilevered wooden 'spring beam' ('spring pole'). Many designs of chisel were tried for different rock formations, and they can be seen in the Zigong museum in Sichuan in China. A chisel weighed

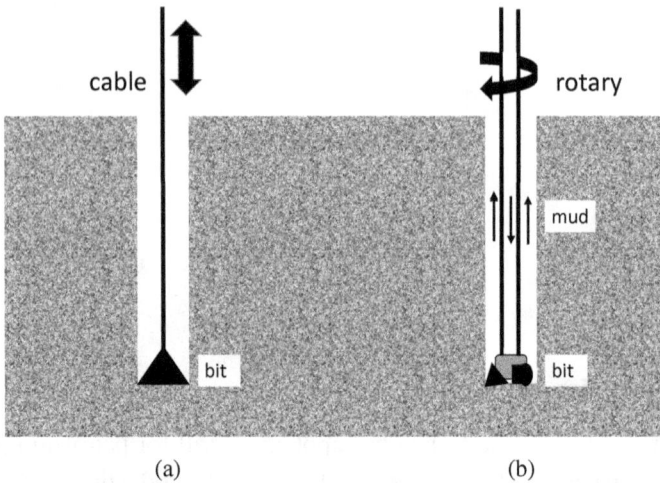

Figure 5.1   Cable drilling (a) and rotary drilling (b)

between 100 and 250 kg. A team of men jumped off the beam to lift the chisel, and then jumped onto the beam so that the chisel fell and broke the rock. The bailer was a length of bamboo, and the one-way valve a strip of ox-hide (Zigong Salt Industry Museum, no date).

This system reached a depth of more than 1000 m a thousand years ago (Zheng and Palmer, 2009). Drilling took a long time: the Deyuan and Sanyuan wells were started in 1889 and completed in 1901. In the US development, men jumping on and off the beam were replaced by horse power and later by a steam engine. The earliest wells in the USA, Europe and Indonesia were all drilled by cable drilling. It is very nearly obsolete in petroleum practice, but is still used for water wells and shallow oil wells.

5.3 Rotary Drilling: Introduction

The principle is illustrated in Figure 5.1(b). The drilling tool rotates: it is either rotated from the surface by a rotating drill pipe or driven at the bottom by a motor. The cuttings are removed by pumping mud down the drill pipe, out through holes in the drill bit, and up in the annular space between the drill pipe and the side of the hole. Rotary drilling was invented in the mid-nineteenth century, initially for water wells: different sources credit inventors in France, Russia and the UK. Lucas used rotary drilling for the famous Spindletop well in Texas and it rapidly supplanted cable drilling.

The drill pipe is rotated by a 'kelly', a square steel hollow bar that fits through a square hole in a rotary table (Figure 5.2): that allows the drill pipe to move downwards while being driven by the rotary table. The drill pipe is a seamless steel pipe with additional thickness ('upset') at either end to allow for the threaded joints ('tool joints') that connect each length to the next. The lower end of each length of pipe is called the 'pin' and the upper end the 'box' (Figure 5.3). The lengths are typically 8.2 to 9.1 m (27 to 30 feet), but longer or shorter lengths can be used. The joints provide a mechanical seal that stops the drilling mud flowing out through the joints. The connected drill pipe lengths are called a 'string'.

At the lower end of the string are heavier and thicker lengths of pipe called 'collars'. Their purpose is to provide weight on the drill bit ('weight on bit' 'WOB') to help it penetrate the rock. During drilling most of the weight of the string is carried by the 'drawworks' at the surface [5.7] so that most of the length is in tension and it does not buckle. Only the lowest part of the string is in compression (Figure 5.4).

5.4 Drill Bits

Below the collars comes the drill bit that cuts the rock to make the hole.

Rotary drilling initially used a 'fishtail' bit, essentially a steel triangle welded to the lower end of the string, not unlike the fishtail bit used in

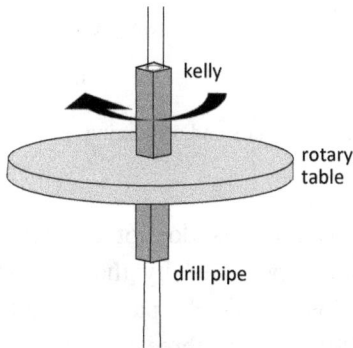

Figure 5.2   Kelly

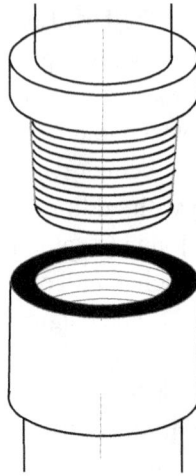

Figure 5.3   Connection between sections of drill pipe

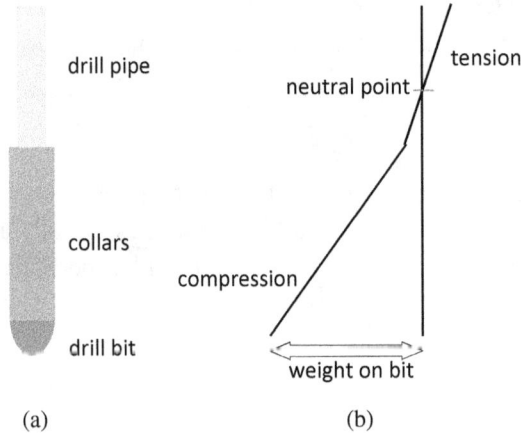

drill pipe

neutral point

tension

collars

compression

drill bit

weight on bit

(a)

(b)

Figure 5.4   Axial force in collars and drill string (a) schematic (b) axial force

cable drilling [5.2]. Fishtail bits do not cut efficiently and wear out quickly, and they were superseded by the roller-cone bit invented by Howard R. Hughes (1869–1924). His patent for a two-cone bit was filed in 1909 (Hughes, 1909), and the three-cone bit ('tri-cone') followed in 1933. Figures 5.5 and 5.6 are two photos of a three-cone bit, and

Figure 5.7 is a schematic looking vertically upwards along the drillpipe axis (a) and horizontally (b).

Each cone carries a number of hardened steel teeth ('cutters') or sintered tungsten carbide inserts, in rows that fit between the teeth on the other two cones. The cones run on bearings in the body of the bit. The axes of the three cones do not always intersect at a point, and then each

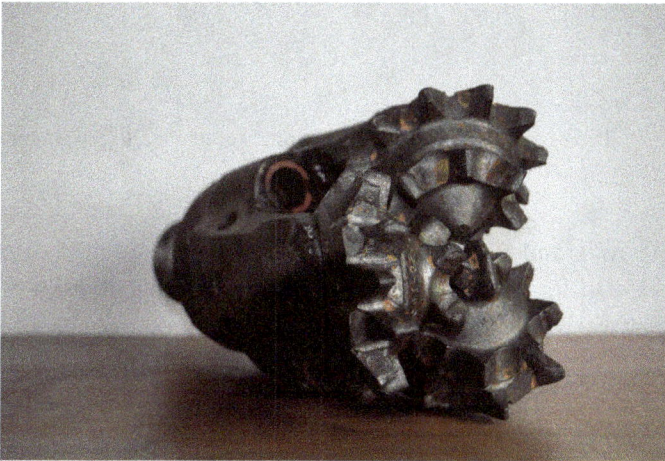

Figure 5.5   Three-cone drilling bit Photo by Jen Flack

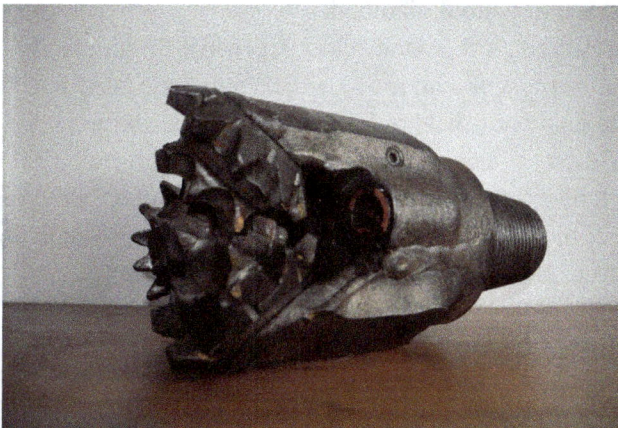

Figure 5.6   Three-cone drilling bit Photo by Jen Flack

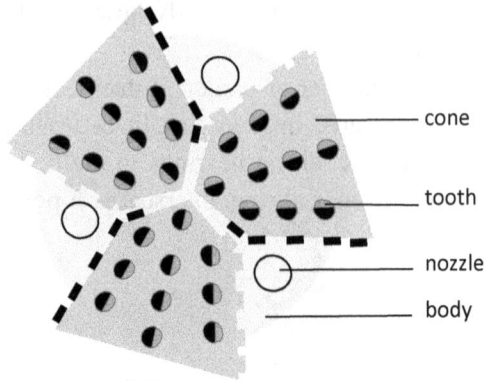

Figure 5.7   Three-cone drilling bit schematic (number of cutters reduced for clarity)

cone both rolls and skids, which strengthens the cutting action in soft formations. Drilling mud [5.5] comes out through three nozzles in the bit. The bit has a threaded connection to a 'bit sub', in turn connected to the lowest collar [5.3], and seals against it. Several kinds of cutters are in use, and are chosen to match the hardness and abrasiveness of the rock to be drilled (Nguyen, 1996; Baker Hughes, 2013).

A second kind of bit does not have rollers, but instead has hard cutters mounted directly on the body of the bit (Figure 5.8). The cutters are synthetic polycrystalline diamond (PDC), thermally stable polycrystalline diamond (TSP) or natural diamond.

Unsurprisingly, a diamond bit is expensive. In an example cited by Nguyen, the net cost of a bit is $41500, the purchase price less the salvage value of the diamonds that are left when the bit is worn out.

The drill bit is plainly a key component of the whole drilling operation. A bit failure has disastrous consequences Careful thought is given to the choice of bit and to the drilling parameters such as weight on bit [5.3], rotation speed, choice of mud [5.5] and mud flow rate.

5.5 Drilling Mud

Mud is pumped down the drill pipe, out through holes in the bit, and up the annular space between the drill pipe and the side of the hole. In the

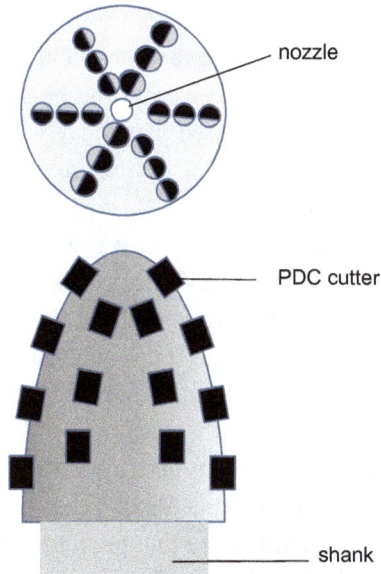

Figure 5.8   PDC Drilling bit schematic (number of cutters reduced for clarity)

early years of drilling, the drilling fluid was water, and later it was a simple mud prepared by trampling clay and water. It is nowadays a much more complex substance, because it has to serve several purposes.

First, the fluid has to clear the cuttings made by the bit and carry them up the hole to the surface, without creating blockages. The cuttings ought not to fall back to the bottom when drilling has to stop, to allow new lengths of drill pipe to be added at the top of the string or to change out a bit.

Second, the fluid has to be heavy enough to apply enough pressure to the sides of the hole, so that in a soft sediment formation the sides are held up and do not fall back. At the same time, we do not want the fluid to flow into the sediment, because then it will be lost. In an extreme case, all the fluid leaks into the formation and none comes back ('lost circulation'). Equally, we do not want the pore fluid already in the sediment to flow out and dilute or contaminate the drilling fluid.

Third, the fluid has to cool the bit, because some of the energy goes into friction and becomes heat, rather than into cutting the formation.

If the bit heats up too much, its steel becomes softer and weaker. The fluid can also reduce friction between the drill pipe and the hole.

Fourthly, the drill bit can be driven at the bottom of the hole rather than the top, by a 'mud motor' placed just above the drill bit. The advantages and broader opportunities of that option are set out in [5.8]. The mud then has to drive the motor.

Fifth and finally, a geologist can make use of the information about the formation that the returning fluid brings back.

Some of those requirements are in conflict. If the fluid is not viscous enough, it will not carry the cuttings. If it is too viscous, it will be difficult to pump. If the cuttings are not to fall back when the flow stops, the fluid has to have some strength even if it is not flowing. The term for that property is 'thixotropy': tomato ketchup and some kinds of paint are thixotropic but water and treacle are not. Ideally, the fluid should plaster the sides of the hole and form a thin cake that prevents inflow or outflow. Too much pressure exerted by the drilling fluid reduces the speed of drilling. If the drilling fluid contaminates the cuttings, the chemical information they carry will be confused. The drilling fluid has to be disposed of.

Much research has been put into drilling fluids. Some muds are water-based and some are oil based. They contain heavy minerals such as barite (barium sulphate) to increase weight, clays to create thixotropy, dispersants to stop the particles clumping together, and granular or fibrous materials to plug permeable zones. A somewhat different fluid is used when the reservoir is reached, because then it is important not to plug the 'pay zone' the production will come from, but at the same time to keep on removing the cuttings.

Back at the surface, the cuttings have to be separated from the mud, either by a 'shale shaker' (a vibrating sieve) or by a 'hydrocyclone' that creates a whirling flow that centrifuges the cuttings out. The cuttings have to be disposed of, and will be contaminated by mud and oil. On some offshore platforms the cuttings have been dumped and left to accumulate under the structure, but that creates other problems.

5.6 Casing and Cementing

The sides of the well cannot be stabilised by drilling mud alone. A section of steel tube ('casing') larger than the drill pipe is lowered down the hole.

Centralisers hold it in the centre, because if the casing touches the hole on one side the cement will not get into the gap. Brushes ('scratchers') remove the mud layer left by the drilling mud. Cement is pumped down the casing, out through a 'casing shoe' at the bottom, and upwards into the annular space between the casing and the side of the well. The length of the cemented section is typically 150 m, but may be more. When the cement has set, the casing and the cement together stabilise the casing, isolate different producing zones, and stop any movement of fluid between the formation and the inside of the casing.

While the well is being drilled, the lowest section of the well is still uncased. When the uncased section is judged to be as long as it can be without creating wall instability or other problems, drilling has to stop and a new section of casing must be set. Each successive length of casing obviously has to be smaller in diameter so that it can go through the preceding sections. There cannot be many casing diameters, because otherwise the smallest would not be able to carry the flow when the well goes into production. Nguyen (1996) discusses the details of casing programmes.

The lowest casing string usually but not always extends to the bottom of the hole and into the reservoir that is to be produced. Flow into the well requires holes in the casing ('perforations'), made by lowering a gun that shoots holes through the casing and into the formation. The next step is a 'drill stem test' ('DST'), which temporarily starts flow through the drill pipe so that the produced fluid's composition, pressure and temperature can be measured. Different zones are isolated by 'packers', which are compressed ring-shaped seals between the drill pipe and the side of the hole.

What can happen when centralisers are left out and cementing goes wrong is described in [8.2.1].

5.7 Drilling Rig

The operations described in [5.3] through [5.6] all require a way of lifting in and out of the well the drilling bit, drill pipe, casing and various other equipment, and of connecting and disconnecting them The drilling rig has to support several other functions, such as the rotation of the drill string, the supply of drilling mud and the separation of cuttings.

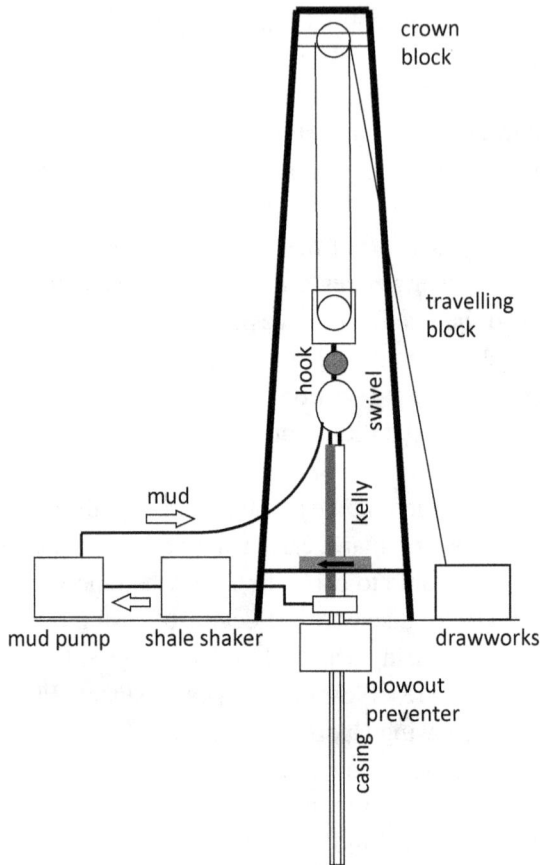

Figure 5.9   Drilling rig (not to scale, and severely simplified for clarity)

Those functions are carried out by the rig illustrated in Figure 5.9, which brings the components together.

The first requirement is to be able to lift and lower things. Lengths of drill pipe and casing are long, and the tower section ('derrick' 'mast') of the rig has to be long enough to allow the lower end of the pipe to be lifted a little way above the rotary table and the kelly. A mobile rig onshore has a mast that can erect itself, so that the rig can be moved. Offshore, on the other hand, the rig moves with the vessel that supports it, and the mast becomes a fixed derrick.

The length of drill pipe lowered at one time is three, four or five lengths ('joints'). Together they form a 'stand'. A longer stand requires a taller derrick, but it saves time during drilling, because fewer connections need to be made up. If the stand is too long it will buckle under its own weight when it is supported at the lower end.

The lengths to be lowered are connected to the 'travelling block'. A wire rope runs from the drawworks (essentially a winch), over the fixed 'crown block' at the top of the derrick, up and down between the travelling block and the crown block, and finally to a deadline anchor. A 'hook' below the travelling block connects it to the swivel and to the drill string below it.

Drilling mud [5.5] has to be able to flow down the drill pipe, and the drill pipe has to be free to rotate. Mud from the mud pump flows into the drill pipe through a swivel, and down through the kelly and the drill-pipe to the drill at the bottom. It circulates back through the annulus and the casing, leaves the rotating string at a second swivel, goes through the shale shaker that separates the cuttings, and returns to the mud pump.

Below the rotary table, the drill string runs through a blowout preventer [5.10].

5.8 Operation of Rotary Drilling Rig

None of the hardware could work without a team of people to operate, manage and maintain it.

Starting with the least skilled and least experienced, the heavy manual work on the rig floor is carried out by 'roustabouts' (floorhands, leasehands). A derrickhand is based at the top of the derrick, and manages the movement of pipe during a 'trip', when the drill pipe is lifted out of the hole so that the bit can be changed out. He also watches the flow of mud, but that task may instead be handled by a mud engineer. The driller is responsible for completing the hole to specification. He sits in a closed cabin ('booth') overlooking the rig floor and has at his fingertips all the information about the progress of the operation. He controls the weight on bit and the rate at which the bit rotates, and decides when casing need to be set. Finally, the toolpusher/rig manager oversees the whole operation.

There are several good videos that describe the different tasks, aimed at people who aspire to work on drilling rigs: see, for example, Righands (2016). In the past drilling rig jobs were essentially closed to women, but that has changed: see, for example, Savanna (2016).

Increasingly, the heavy manual tasks are being mechanised. For instance, stands of pipe have to be threaded together and tightened when a new stand is added. That used to be done by wrapping a chain around the drillpipe and tensioning the chain with an air tugger, a slow operation which is potentially dangerous if the roustabouts are not careful. An 'iron roughneck' carries out the same task faster and more safely.

YouTube has accessible videos of drilling.

5.9 Directional, Horizontal and Multilateral Drilling

Wells were originally drilled vertically. The ancient Chinese cable-drilling system [5.2] attached high importance to keeping the well vertical and correcting it if it shifted out of vertical. Later it came to be realised that there could be advantages in drilling out of vertical ('slant drilling'): in particular it allowed a driller occupying one piece of land to drill surreptitiously under the boundary between his land and the next owner's and to steal his oil. If the extent of the reservoir was known, slant drilling could site the drilling operation a little distance away, so that it could avoid a town, a national park or a historic site. The same principle can be used to drill wells in different directions from the same wellsite. That option is particularly valuable offshore, because it eliminates the need to have a different platform for every well. If a well has to be abandoned, for example if the drilling string breaks, directional drilling can be used to bypass the obstruction ('sidetracking'). Several 'multilaterals' going in different directions can be drilled from the same well, by making a hole in the casing of the first well and deviating to a different direction or a different level. Finally, if a well has blown out [5.9] slant drilling can drill a relief well to the same reservoir, so that its pressure can be reduced and the first well brought back under control.

One option is to tilt the whole rig and to drill a straight hole from the tilted mast, but that severely limits the options. The tool first used to deviate a well downhole was a 'whipstock', a wedge-shaped trough

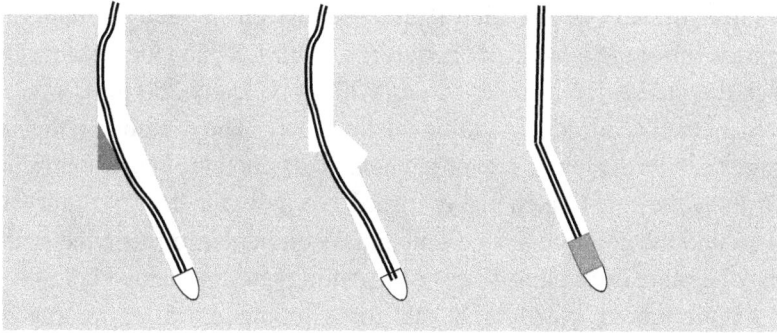

Figure 5.10   Directional drilling alternatives (not to scale, and angles exaggerated for clarity)

lowered into the well, with the desired orientation (Figure 5.10(a)) and at the 'kick-off point' ('KOP') where the deviation from the vertical is to start. The whipstock diverts the drill bit mechanically. An alternative is to jet a cavity on one side of the well (Figure 5.10(b)), and the drill bit moves into the cavity and continues in the same direction. The driller of course wants to confirm the slope and the orientation of the well, and he has devices to do that, based either on a pendulum that uses gravity as a reference direction or on a gyroscope that uses the Earth's rotation (Nguyen, 1996).

Directional drilling today deviates the well by the scheme shown in Figure 5.10(c). The drill bit is rotated by a mud motor [5.5] at the lower end of the drilling string. The flow of drilling mud through the motor creates a torque on a helical rotor, and that torque is balanced by torque in the string. Above the motor is a 'bent sub', as its name suggests a short length of bent pipe. Alternatively the bend can be combined with the motor ('bent housing'). The direction of the deviated hole has to lie in the plane of the bent sub, but the orientation or inclination of the hole can be altered by rotating the bent sub from the surface. Once the intended direction has been reached, drilling can revert to the conventional scheme with a rotating drillstring. In this way it has become possible to drill geometrically complex hole paths to intersect different strata.

The bent sub ought not to have a sharp bend, because then the drillstring would be overstressed, particularly if it were later rotated so that the bending stresses would alternate between tensile and compressive. The bent sub angle is typically between 0.5 and 1.5°, and the distance over which the inclination changes through 70° is typically 250 m.

Slant wells introduce additional problems. The weight of the drill strings rests on the lower side of the hole, and generates increased friction, both along the hole (which makes it harder to move the drillstring forward) and around the circumference (which makes it harder to rotate the string). It may be more difficult to remove the cuttings from a slanted hole.

At one time it was thought that those problems would become progressively more severe as the angle to the vertical increased, and that it would not be practicable for the angle to exceed about 60°. Later it came to be realised that some of the problems actually became smaller, and that wells could be drilled horizontally or even upwards.

Figure 5.11 is a sketch. The topmost section of the well is vertical. It then curves round to a slant section, and then curves again to a section with a near-horizontal direction. If a relatively thin reservoir is to be drilled, that section can follow the reservoir. It can been seen at once that the length of the well that intersects the reservoir is much greater than it would be if the well had been drilled vertically. That greater length can be perforated, and petroleum can flow into the well all the way along, and that much increases production.

Horizontal drilling is nowadays routine, and indeed some would argue that the purely vertical well is obsolete. It is one of the two key factors that

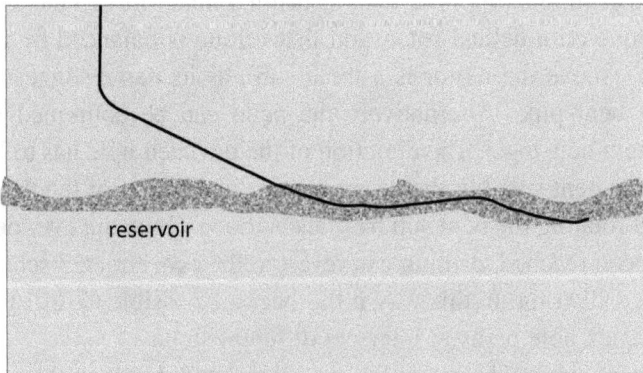

reservoir

Figure 5.11 Horizontal drilling (not to scale)

have made possible the rapid development of shale gas and shale oil. The second factor is hydraulic fracturing ('fracking'), which applies a high pressure to the sides of the hole and cracks the rock, so that more gas and oil can flow into the hole.

5.10 Blowout Prevention

A well can get out of control. That does not often happen nowadays, because of better design of wells and more precise procedures. If it does happen the consequences are extremely serious [8.2]. Oil and gas spew out ('blow out') and cause fire, explosions and environmental damage. The casing and production tubing can be blown out of the well.

A blowout preventer ('BHP') is a system of valves and shear rams that is fixed to the topmost section of casing ('conductor'). If the driller concludes that a blowout is beginning to happen, the valves shut and the shear rams close off the pipe.

5.11 Fishing

Things can fall down the well, if for example a drill bit breaks and roller cones fall out, or if the drill string fails in torsion. It may not be possible to drill through them, and it is disastrous if the obstruction prevents the well from being completed.

One option is to divert the well around the obstruction ('sidetracking'), applying the techniques described in [5.9]. Alternatively, the broken fragments can be removed from the well, a process called 'fishing' (Nguyen, 1996). Much ingenuity has been applied, starting with the Chinese wells described in [5.2], which applied various barbed hooks and spears (Zigong Museum, no date). One modern scheme is the 'junk basket': the basket is placed above the bit, mud is circulated to lift the fragments, and then the flow is suddenly stopped and the fragments fall into the basket. Another scheme uses magnets and another a 'junk mill' that breaks up metal fragments.

References

Baker Hughes. Drill bits catalog. online at www.bakerhughes.com (2013).

Brantly, J. History of oil well drilling. Gulf Publishing (MORE).

Hughes, H.R. US patents 930758 and 930759 (1909).

Nguyen, J.P. Drilling. Editions Technip (1996).

Pees, S.T. Oil history. www.petroleumhistory.org/OilHistory (2015).

RigHands. www.righands.com (2016).

Savanna. Video 'Dee — female rig hand' www.savanna.com (2016).

Sun, J.S. A city with salt wells everywhere. Guangxi Normal University Press (2005).

Zheng, J. and Palmer, A.C. Bamboo pipelines in ancient China — and now? *Journal of Pipeline Technology*, **8**, 95–98 (2009).

Zigong Salt Industry Museum (in collaboration with Shell). Drilling and gas recovery technology in ancient China. Zigong Salt Industry Museum, Zigong, Sichuan, China (no date).

Chapter Six

Producing Petroleum on Land

6.1 Introduction

Oil and gas are found in porous rocks, sometimes close to the ground surface but usually much deeper, at a depth between 500 and 10000 m. A typical pore diameter is 0.1 mm. If it is to be useful, the petroleum has be brought to the surface. This is done by drilling into the geological reservoir structure where the petroleum is trapped, and creating a pathway it can flow through.

Often the petroleum in place in the reservoir is under high pressure. When we drill into the reservoir, we can reduce the pressure at the hole, and petroleum will then flow from the reservoir and into the hole. If the reservoir pressure is high enough, the petroleum reaches the ground surface. If the flow is not contained and controlled, the petroleum continues to flow upward into the air, creating a fountain ('gusher'), a situation we want to avoid, because it is wasteful, environmentally damaging and a fire risk. Instead we want the flow to go into a pipe, where the flow can be controlled by valves and if necessary stopped.

If the reservoir pressure is not quite so high, it may not be enough to drive the petroleum all the way to the surface, against the pressure created by the unit weight of the petroleum multiplied by the height difference between the reservoir and the surface. It may then be necessary to increase the drive pressure or to reduce the pressure difference. One option is to install a downhole pump at the bottom of the hole, to create an additional pressure that will help to drive the petroleum towards the surface. Another option is to inject water under pressure into a reservoir in additional injection wells, so that the high pressure around the injection wells pushes oil towards a separate producer well at a lower pressure. Another option is to reduce the pressure difference created by the weight of petroleum in the hole, which we can do by injecting compressed gas near the bottom of the hole.

One decision we have to make is how many wells we need to drill. If we try to produce from a large reservoir with just one well, then once the petroleum in the immediate vicinity of the well has been produced, the rest will have to flow a long way, and so we can expect production to be slow. If production is very slow indeed, it may not be economic to continue to produce at all, and so we may need to stop production and abandon the well, even though the reservoir still contains substantial volumes of petroleum. If instead we drill many wells, production will be faster, but the wells will cost much more. Somewhere in between there is an optimum that allows continuing production at a reasonable rate but keeps drilling costs within bounds.

The rate at which petroleum flows towards a well is sensitive to the permeability of the rock, a concept that will be quantified in section 6.2. It may be possible to increase the effective permeability by breaking the rock in place ('fracturing', 'fracking'), but that too costs money and may create other difficulties. Fracking is discussed in section 6.5 below.

All these questions and many more involve fluid flow through rock. Section 6.2 is concerned with flow of a single fluid phase, oil or gas or water, but not more than one at a time. Section 6.3 deals with multiphase flow, where more than one fluid is present and can flow. Section 6.4 is concerned with the effect of interfacial tension ('surface tension), and 6.5 with fracking. Section 6.6 deals with pressure maintenance, and 6.7 with well spacing.

6.2 Oil and Gas Flow in Rocks; Single-Phase Flow

We need to be able to describe and quantify the flow. Oil flows along tortuous paths through pores in rock and gaps between the particles. Figure 6.1 below is obviously two-dimensional, but the true structure is highly three-dimensional, and contains pores and particles with a wide range of sizes and shapes.

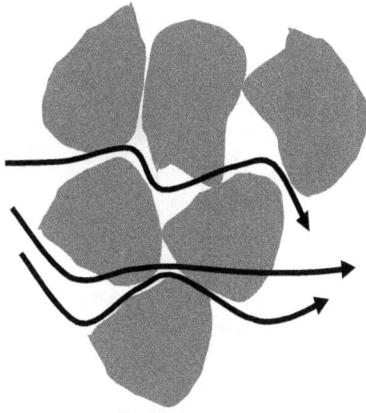

Figure 6.1   Flow through a porous rock

How easily the oil flows depends on the size and shape of the pore spaces, on whether they are blocked by other fluids such as water and gas, and on the viscosity of the oil.

Consider first a heavily-idealised model of an element of an isotropic porous rock, and steady flow of an incompressible Newtonian fluid. Isotropic means that the rock has the same properties for all directions of flow. Steady means that the pressure distribution and flow are independent of time. Incompressible means that the fluid density is independent of the pressure. Newtonian means that the shear strain rate is proportional to the shear stress.

The element has dimensions dx, dy and dz. The pressure changes in the x-direction, and the pressure is p at the left-hand side and $p + dp$ at the right. The volumetric flow is in the same direction at the pressure gradient, and is q per unit area per unit time; the area is the total area perpendicular to the flow, not just the area of the tubes. The fluid flows through circular tubes, all with the same radius a, and all parallel to the x-axis. The number of channels per unit area is n.

If we put two of these blocks side by side (Figure 6.3), the area is multiplied by 2 and the flow rate through the two blocks together is multiplied by 2, but the pressures are unchanged:

If we put two blocks end to end in the flow direction (Figure 6.4), and make the flow rate the same as in Figure 6.2 the pressure change is from p to $p + dp$ in the nearer block, and from $p + dp$ to $p + 2dp$ in the further block.

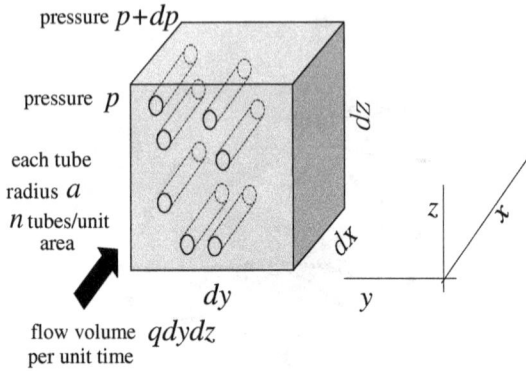

Figure 6.2   An idealised model of 1-dimensional flow in a porous rock

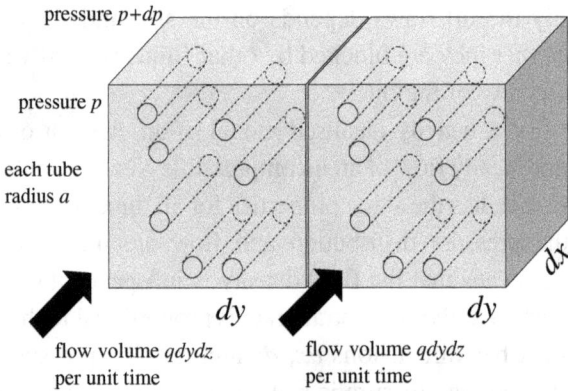

Figure 6.3   Two of the blocks in Figure 6.2 put side-by-side

Volumetric flow rate q is the volume of fluid per unit time per unit area at right angles to the flow, counting the total area, not just the tubes. The incomplete preliminary analysis above suggests to us that the volumetric flow rate q depends on the pressure gradient dp/dx in the flow direction. It will also depend on the radius a of each tube (because experience suggests to us that the smaller the tube, the more difficult it is for a viscous fluid to flow through it, just as honey will flow more rapidly out of a bottle with a wide neck than out of a narrow neck), on the number of tubes n per unit area (because the more tubes there are, the more flow

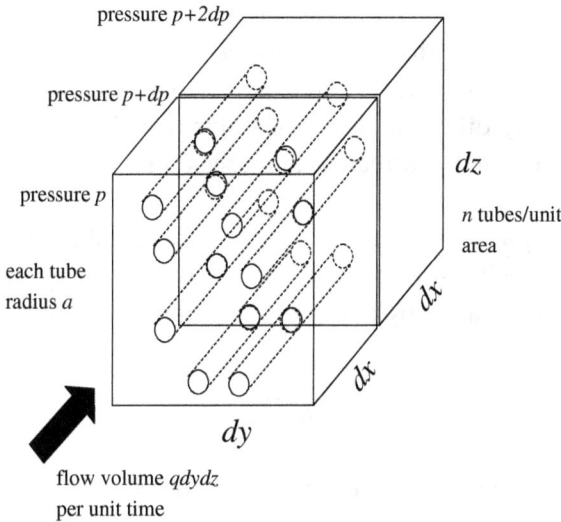

Figure 6.4   Two of the blocks in Figure 6.2 put end-to-end in the flow direction

there will be), and on the viscosity μ of the fluid (because we would expect a viscous fluid like honey to flow more slowly than a less viscous fluid like water, other things being the same).

We could then carry through the classical Hagen-Poiseuille analysis of flow in circular tubes (found in any fluid mechanics text), but it is easier to use dimensional analysis (see, for example, Palmer (2008)). Dimensional analysis tells us that a properly-constituted relationship must be expressible as a relationship between dimensionless groups.

The dimensions of q and the quantities it depends on are

q	(volume/area)/time	$[(L^3/L^2)/T] = [LT^{-1}]$
a	radius of one tube	$[L]$
n	(number of tubes/ area)	$[0/L^2] = [L^{-2}]$
dp/dx	(pressure/distance)	$\dfrac{\text{force/area}}{\text{distance}} = \dfrac{[MLT^{-2}]/[L^2]}{[L]} = [ML^{-2}T^{-2}]$
μ	viscosity	$\dfrac{\text{stress}}{\text{velocity gradient}} = \dfrac{[MLT^{-2}]/[L^2]}{[LT^{-1}]/[L]} = [ML^{-1}T^{-2}]$

where the fundamental dimensions have been taken as mass, length L and time T, because that is the conventional choice. There is nothing magic about that choice of dimensions: Palmer (2008) shows by example that any different choice of fundamental dimensions leads to the same result, and the reader can check that the same conclusion applies here.

We expect q to be proportional to n, because the tubes are independent, and if there are twice as many tubes there is twice as much flow. We start the dimensional analysis by looking for a group of the form

$$\frac{q}{na^{\alpha}(dp/dx)^{\beta}\mu^{\gamma}}$$

where α, β and γ are not yet known. Its dimension is

$$\frac{LT^{-1}}{[L^{-2}][L]^{a}[ML^{-2}T^{-2}]^{\beta}[ML^{-1}T^{-1}]^{\gamma}} = [M]^{-\beta-\gamma}[L]^{3-\alpha+2\beta+\gamma}[T]^{-1+2\beta+\gamma}$$

If the group is dimensionless, the powers of M L and T must all be zero, and so

[M]	$0 = -\beta - \gamma$
[L]	$0 = 3 - \alpha + 2\beta + \gamma$
[T]	$0 = -1 + 2\beta + \gamma$

Those three simultaneous equations can be solved by adding twice the first equation to the third, which gives γ, substituting into the third equation to get β, and substituting into the second equation to get α. The results are $\alpha = 4$, $\beta = 1$ and $\gamma = -1$, and the dimensionless group is

$$\frac{q}{na^{4}(dp/dx)\,\mu^{-1}}$$

If no other parameters are involved, there is nothing else for that dimensionless group to be a function of. It must be an unknown constant c, and so

$$q = \frac{cna^{4}}{\mu}\frac{dp}{dx} \tag{6.1}$$

The flow rate is proportional to pressure gradient and inversely proportional to viscosity, as we would expect. cna^4 is a property of the internal geometry of the porous rock, and we call it permeability, denoted k, so

$$q = -\frac{k}{\mu}\frac{dp}{dx} \qquad (6.2)$$

The minus sign is put in to keep the signs straight, because we want to count as positive a flow in the positive x-direction. If dp/dx is negative, the pressure decreases in the x-direction, and the flow is down the pressure gradient from high pressure to low pressure and therefore in the positive x-direction.

The permeability k is independent of the properties of the fluid. Permeability has dimension (length)^2. As you can see from the presence of the a^4 term in cna^4, permeability of a porous rock depends very strongly on the sizes of the pores.

From the argument above, permeability ought ideally to be defined as (length)^2. However, for historical reasons that it is now too late to change, petroleum engineers almost always take the practical unit of permeability as the Darcy (D), a metric but not SI unit called after Henri Darcy (1803–1858), who did famous work on the water supply to Paris. A rock has a permeability of 1 D if a pressure gradient of 1 atmosphere/cm induces a flow rate q of 1 cm/s in a fluid of viscosity 1 cP (centipoise). 1 centipoise is another metric but non-SI unit, and is 0.001 Pa s in SI units. 1 D is about 10^{-12} m^2. It is rather a large unit and the more useful unit is the milliDarcy, mD, equal to 0.001 Darcy and therefore 10^{-15} m^2. Reservoir permeabilities usually lie between 1 and 500 mD.

If you have studied geotechnics, note that what geotechnics calls permeability is qualitatively different from the definition above, and is defined as k/μ rather than k. The reason is that in geotechnics the fluid is almost always water, and variations in water viscosity are ignored (even though variations with temperature are actually quite significant). In this book, we shall only use the petroleum definition in equation (6.2) and not the geotechnics definition.

Going back to equations (6.1) and (6.2), we can see that the permeability cna^4 depends strongly on the tube diameter a. If we increase the diameter by a factor of 10, we increase the permeability by a factor of 10,000. If we have a more complex structure that includes tubes with different diameters, and if we scale all the linear dimensions up by a factor of 10, again we increase the permeability by a factor of 10,000. Together that tells us that the permeability of a reservoir is strongly dependent on the dimensions and number of the largest pores. In terms of actions that an engineer might take, it tells us that if we can do something to increase the number and size of the largest pores, for example by creating cracks, that ought to create a strong and useful increase in permeability. That possibility is examined further in section 6.5.

The argument above tacitly assumed that viscosity was the only parameter needed to describe the fluid. Hydraulic analysis of flow in pipes finds that at larger velocities turbulent flow occurs when the Reynolds number Re exceeds about 2000, and that then the fluid density becomes important. In a circular pipe carrying a Newtonian fluid

$$\text{Re} = \frac{\rho U D}{\mu} \tag{6.3}$$

where ρ is fluid density, U is mean velocity averaged over the pipe cross-section, D is internal diameter, and μ is viscosity as before. We therefore need to check whether or not turbulent flow might be important. Consider a horizontal flow of 20,000 b/d into a well over a payzone 8 m thick. At a radius of 10 m from the well axis, the velocity averaged over the total area is

$$q = \frac{20000 \text{ b/d} \times 0.159 \text{ m}^3/\text{b} \times \dfrac{1 \text{ day}}{86400 \text{ s}}}{2\pi(10)(8)} = 7.3 \times 10^{-5} \text{ m/s} \tag{6.4}$$

If the porosity is 0.1, a typical velocity in a pore might be 10 times larger, say 7.3×10^{-4} m/s. If the diameter of a typical pore is 10^{-4} m, the oil density is 800 kg/m^3, and the oil viscosity is 0.01 Pa s (typical values), then

$$\text{Re} = \frac{(800)(7.3 \times 10^{-4})(10^{-4})}{0.01} = 0.006 \tag{6.5}$$

a very long way smaller than the critical Re value of 2000 at which there is a transition from laminar to turbulent flow in a pipe. This is a rough calculation, but any sensible variation of the parameters will still lead to a very small value of Re. It follows that turbulent flow can only occur very close to the wellbore, for instance in a zone with large particles, such as a gravel pack immediately outside the bore: we do not normally need to consider it.

The introduction above assumed the flow to be one-dimensional in the x-direction. In general the pressure will vary in all three coordinate directions x y and z. The flow per unit area in the x-direction is q_x per unit area, and the corresponding flows in the y and z directions are q_y and q_z.

An isotropic material has the same permeability in every direction, and then we can generalise equation (6.2) to

$$q_y = -\frac{k}{\mu}\frac{\partial p}{\partial y}$$

$$q_z = -\frac{k}{\mu}\frac{\partial p}{\partial z} \tag{6.6}$$

$$q_x = -\frac{k}{\mu}\frac{\partial p}{\partial x}$$

Most geological materials are not isotropic. Imagine a structure with alternate horizontal layers of sandstone and shale. (Figure 6.5).

Each sandstone layer is 2 m thick and has permeability 500 mD. Each shale layer is 0.2 m thick and has permeability 0.5 mD. The shale layers are much thinner than the sandstone layers, but have a far smaller permeability. The horizontal permeability is only a little less than 500 mD, because a fluid can easily flow horizontally through the sandstone but cannot easily flow through the shale. The vertical permeability is much smaller, because a fluid flowing vertically has to go through the shale. Almost all the pressure drop is in the shale, and the vertical permeability is a little more than $0.5 \times (2.2/0.2) = 5.5$ mD. This is a simple idealisation of the kind of structure seen in Figure 6.6, which is a photograph of a cliff near Bridport in Dorset, England, not far from the Wytch Farm oilfield.

A further complication is that the dominant directions in the structure may not coincide with the coordinate directions. Think of the Figure 6.5 structure tilted in Figure 6.7(a), or tilted and ruptured by a fault in Figure 6.7 (b):

On a small scale, entirely within either a sandstone layer or a shale layer, and assuming each layer to be isotropic, the flow is in the same

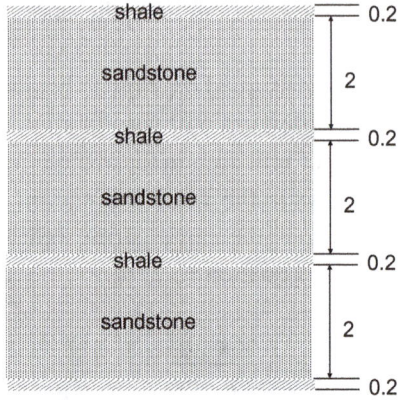

Figure 6.5 Alternating horizontal layers of high-permeability sandstone and low-permeability shale

Figure 6.6   Sea cliff on south coast of England near Wytch Farm oilfield photo
Source: Chris Downer geograph.org.uk.

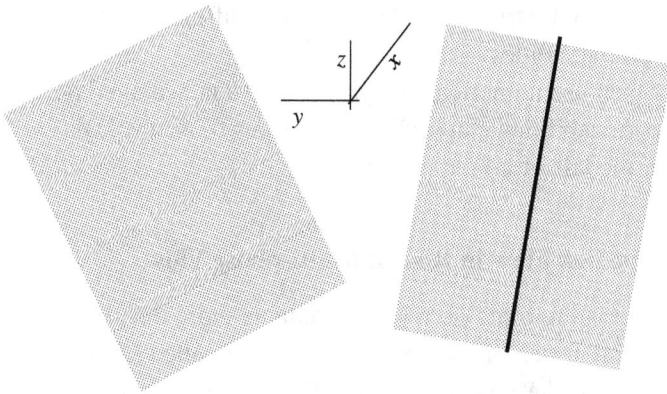

Figure 6.7   Structure (a) tilted (b) tilted and cut by a fault

Figure 6.8   Model of complex structure in which different sections have different permeabilities

direction as the pressure gradient, and is governed by (6.6). On a larger scale, though, a pressure gradient in the x-direction induces a flow in both the x- and z-directions, and vice versa.

Real structures are far more complex than these idealisations. Reservoir engineers use geological knowledge and data from boreholes to synthesise model structures that include different rocks with vastly different permeabilities:

They then use numerical analysis to arrive at averaged horizontal and vertical permeabilities ('upscaling'), and put those averaged values into numerical models of a field as a whole.

6.3 Oil and Gas Flow in Rocks: Multi-phase Flow

More often than not, the pore space contains more than one fluid, so that oil, water, and gas are all present. A parameter called saturation describes what fraction of the pore volume is filled by each phase. The fraction of the pore volume occupied by oil is the oil saturation, and similarly for gas and water. Thus, if oil fills three-fifths of the pore volume, gas fills three-tenths of the pore volume, and water fills the remaining one-tenth, the oil saturation is 0.6, the gas saturation 0.3 and the water saturation 0.1.

Each phase partially or completely obstructs the movement of the other phases. That can be modelled with a concept called relative permeability.

The simplest possibility is a one-dimensional flow in a two-phase oil-water system. Equation (6.2) is generalised by adding a multiplier $k_{rel,o}$ called relative permeability, so that

$$q_o = -\frac{kk_{rel,o}}{\mu_o} \frac{dp}{dx} \tag{6.7}$$

where q_o is the volumetric flow rate for oil, k is the permeability in a single-phase flow, μ_o is the oil viscosity, and $k_{rel,o}$ is the relative permeability for oil. $k_{rel,o}$ is dimensionless, and is a measure of how much the flow of one fluid is obstructed by the presence of the other fluid. If the pores are completely filled with oil, corresponding to an oil saturation of 1 and a water saturation of 0, the relative permeability for oil $k_{rel,o}$ is 1 and the corresponding relative permeability for water $k_{rel,w}$ is 0. If the oil saturation is less than 1, so that the pores are only partly occupied by oil and the rest of the pore space is filled with water, $k_{rel,o}$ is less than 1. In general, for the flow of water

$$q_w = -\frac{kk_{rel,w}}{\mu_w} \frac{dp}{dx} \tag{6.8}$$

where q_w is the volumetric flow rate for water, μ_ω is the water viscosity, and $k_{rel,w}$ is the relative permeability for water. The relative permeabilities $k_{rel,o}$ and $k_{rel,w}$ do not need to add up to 1, and their sum is always less than 1, and usually much less.

The relationship between the relative permeabilities and the oil and water saturations is shown schematically below:

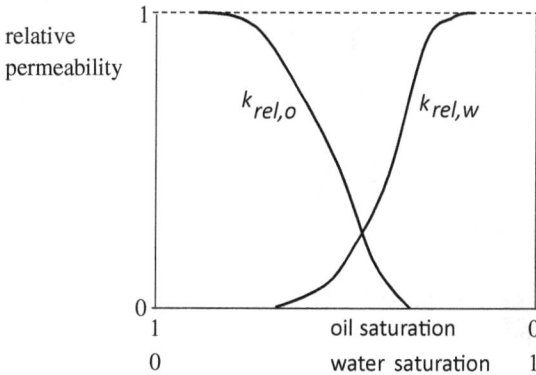

Figure 6.9   Relative permeabilities in two-phase flow

Notice that if the oil saturation falls below a certain value, about 0.4 in the schematic, $k_{rel,o}$ is zero and the oil 'sticks' and does not move at all.

The analysis can easily be extended to three or more phases. A further complication is that the relationship depends on the direction of flow: it is not the same for water displacing oil (the common case) as it would be for oil displacing water.

6.4 Interfacial Tension

If you lower a clean small-diameter glass tube into water, the water wets the inner surface of the tube and climbs a little way up it, as in Figure 6.10(a). This is a capillarity effect of surface tension at the interface between water and air. The effect is familiar from daily experience. It is barely noticeable if the tube is large (for example a test tube) and much greater if the tube is larger, for example a fine capillary, or the pore

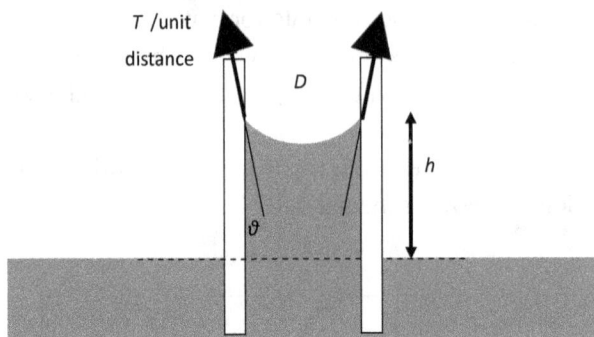

Figure 6.10   Capillary forces in a vertical tube

volumes in fine sand or blotting paper. Surface tension plays an important part in the movement of water within soils and in the formation of droplets.

The effect can be quantified by the simple model of a vertical tube in Figure 6.10. The tube is circular with diameter D, the water has density ρ, and the gravitational acceleration is g. The surface tension can be treated as a tensile force in the surface, T per unit length. At the point where the air, the water and the glass meet, the water meets the glass at a small angle θ rather than tangentially. Writing down the equation expressing vertical equilibrium of the water above the dashed line (drawn level with the surface outside the tube), and taking D as small compared to h

$$T\pi D\cos\theta = (\pi D^2/4)h\rho g \tag{6.9}$$

and so

$$h = \frac{4T}{\rho g D}\cos\theta \tag{6.10}$$

For water and air, T is 0.07 N/m. Taking ρ as 1000 kg/m³, g as 9.81 m/s², θ as 20°, then for a 1 mm (0.001 m) diameter tube, h is 0.027 m (27 mm); you can check this for yourself if you can find a small-diameter glass tube. The pressure difference between the inside and the outside of the tube at the level of the dashed line is 260 Pa. Even if the tube diameter were 0.1 mm, smaller by a factor of 10, that pressure difference would be only 2.6 kPa, about one-fortieth of atmospheric pressure and small by comparison with typical pressure differences within an oil reservoir. If the diameter of the pore were much smaller, however, the pressure difference needed to drive the water through the pore would be correspondingly larger. Moreover, if a number of interfaces are in series, the pressure differences add up, because a certain pressure difference is required to drive each interface through a gap between particles, and so the overall effect is multiplied. In an oil/water system, the interfacial tension is somewhat smaller than for water and air.

When we are injecting water into an oil-bearing formation to try to get the water to displace oil, we can expect to increase the production of oil if we can reduce interfacial tensions between oil and water. We may be able to do that by adding a surfactant to the injection water (Reisberg, 1967). The surfactant can reduce the interfacial tension to as little as 10^{-6} N/m, and the effect of interfacial tension is then negligible even for very small pores.

6.5 Fracturing

The analysis in section 6.2 demonstrated how strongly the permeability depends on the size and number of the pore volumes between the particles. If we can somehow intervene to increase those pore volumes, we can expect much to increase the effective permeability of the reservoir. We will not need to increase all the pore volumes: if we can create an irregular network of fine cracks in the rock, and if we can keep those cracks open, the oil or gas will flow a short distance through the rock to the nearest crack and then flow onwards through the cracks.

Even modest cracks can be highly effective. Hunt (1996) quotes a good example cited at an AAPG meeting. He tells us to think of a 1 mile (1.61 km) square layer of 10^{-8} D (0.01 μD) permeability shale in an oil reservoir, and to imagine that shale layer crossed by a single straight crack from one corner to the diametrally opposite corner. He asks how wide the crack has to be before as much oil flows through the crack as flows as through the remainder of the shale layer. The answer is 6 microns (0.000006 m). (If it worries you that this answer is independent of both the thickness of the layer and the viscosity of the oil, you ought to reread section 6.2).

Over many tens or hundreds of millions of years, petroleum moved from the source rock where it was formed to the reservoir rock where it is now. The reservoir rocks that petroleum engineers are principally concerned with have permeabilities between 1 and 500 mD. Generally, the source rock and the reservoir rock are not the same. The permeability of the source rock is very much smaller, between 1 mD and 10^{-8} mD. Shale, for example, is a fine-grained sediment deposited by slow-moving water, a mixture of clay minerals, kaolinite, montmorillonite and illite (the dominant clay minerals found in clay in other contexts), together with larger particles of other minerals such as quartz and calcite. In some contexts shale's low permeability lets it function as a cap rock that holds oil and gas within a reservoir.

Shale is a potential source of oil and gas, and then it functions both as the source rock and a reservoir rock, but its in-situ permeability is so low that the oil and gas currently within it cannot be produced conventionally. The shale can be mined in the same way that coal is mined, either underground or in an open-cast pit. The shale is then transported to a power station and burned like coal. That is occasionally done, notably in Estonia where mined shale generates 80 per cent of the nation's electricity, but it is generally economically unattractive by comparison with conventional oil and gas. An alternative is to heat the coal to 450°C, and the kerogen then breaks down into oil, gas and a solid residuum, which can be refined further or burned as they are.

Over the last twenty years, another option has become extremely popular. Water is pumped down a drilled hole, and applies a high hydraulic pressure to the hole. The high pressure creates a high circumferential tensile stress, and the rock breaks in tension, creating a network of fine cracks around the hole. Fragments of a strong solid material called a proppant, quartz sand, aluminium oxide or something similar, are pumped down the hole with the water. They are carried by the water into the newly-created cracks, so that when the pressure is later reduced the proppant holds the cracks open. Depending on the applied pressure and on the initial stress state in the rock, the fractures can extend for some distance. Once the pressure has been removed, gas can flow into the well. Fracturing also happens naturally. Many rocks include thin veins of a visibly different material, and those veins were created by fracturing.

The water injected in hydraulic fracturing is not usually pure water. The engineer can incorporate additives to accomplish the objectives better, and can change the composition as the fracturing proceeds. The most commonly applied additives are gels, either guar gum or cellulose derivatives such as carboxymethyl cellulose (which increase the viscosity and help the fluid to carry the proppant, and form a gel), borax and boric acid (which further increase the viscosity), metal compounds (which cross-link the gel), and later in the operation enzymes and oxidisers that break down the gel so that it can flow back out of the fractures and does not carry the proppant back. Other additives include acids (which help initiate fractures), corrosion inhibitors and friction reducers (which reduce the pressure drop between the surface and the fractures). The proppant can incorporate a radionucleide tracer so that movement of the proppant can be monitored

The underlying idea behind fracturing is far from new. Soon after oil production in the US began in 1859, engineers experimented with exploding dynamite or nitroglycerine at the bottom of wells, and a patent for an 'exploding torpedo' was applied for in 1865. Those fractures cannot have extended far, but an increase of production was observed. The technique was later applied to water and gas wells. The modern idea of fracturing with water pressure was first tried out experimentally in 1947, and its commercial application began two years later. It has now been used to stimulate a million oil and gas wells in the US (Gallegos and Varela, 2014). Its recent application to shale gas started in 1973 and is now having an impact across the world petroleum industry. Estimated gas reserves in the US total 2300 Tcf (6.5×10^{15} m^3), and a quarter of that is in shale. In 2001 only 1.6 per cent of US natural gas was produced from shale, but by 2010 more than 20 per cent came from that source, and the US Energy Information Administration estimates that in 2035 the proportion will be 46 per cent. Active development of shale gas is now increasingly extending to other countries, China, Norway, Brazil, Poland and many others.

Hydraulic fracturing is often combined with horizontal drilling, discussed in section [5.9].

The application of hydraulic fracturing to shale gas and oil is an exciting development. At the same time it is deeply controversial and the subject of a lot of misrepresentation. It has been grabbed as an issue by people opposed to the petroleum industry for quite different reasons, though that does not noticeably discourage them from owning cars or using electricity. On the day this is written, a small rural community nearby is torn by controversy after 500 people came to a meeting to vote on an ordinance that would ban fracturing.

There are genuine concerns. Hydraulic fracturing uses a great quantity of water, and not all of it is recovered. An average well uses 11,000 to 30,000 m³ (3 to 8 million US gallons) over its lifetime, but according to one report, in the US the volume of water used in hydraulic fracturing is 1.3 per cent of the volume used in car washes. Additive chemicals help the water to flow easily, but can contaminate ground water and creates a disposal problem when some of the water subsequently returns to the ground surface. Deliberately, hydraulic fracturing alters the state of stress in the rock. In a seismic area, that addition to the pre-existing stress state may be enough to trigger earth tremors. Almost all those tremors are small, below the limit at which human beings can detect them, but under the right conditions a larger earthquake is possible. A careful and balanced article by Rubinstein and Mahani (2015) cites two magnitude 4.4 earthquakes in Alberta and British Columbia to 2012. According to another article, there are four instances where hydraulic fracturing has induced seismic events large enough to be felt by people, one in the US, one in Canada and two in the UK. The United States Geological Survey publications (2005) have a lot of material on earthquakes and fracturing, but it does not support the opinion that it is a major problem; see also Bame and Fehler (1986) and Kim (2013).

A few studies of hydraulic fracturing suggest that the effect is to create a short-lived boom, whose benefits are unevenly distributed across the community and are not sustainable. It has that in common with many if not most developments of natural resources. Against that, the industry is pursuing hydraulic fracturing with great energy, and it is having a major economic impact.

6.6 Pressure Maintenance

Once oil begins to be produced from a field, the pressure in the reservoir tends to drop. Gas is usually dissolved in the oil, and a reduction of pressure leads gas to come out of solution and form bubbles within the oil (just as when you remove the cap from a bottle of fizzy drink, the pressure inside drops to equal the atmospheric pressure outside, and the reduction of pressure allows gas to come out of solution as small bubbles and to fizz up). Some of the gas will rise into a gas cap above the oil, and the gas cap will expand in volume. Some of the gas will remain in the oil reservoir, where it will partially obstruct the pores and reduce the flow (in the way described in section 6.3).

In the early years of oil production, only the oil was valued, whereas the gas was no more than a nuisance and was flared. The pressure in the field dropped, and ultimately it fell to a level at which it was too small to drive oil to the ground surface. The field was then abandoned. It was recognised that this strategy left a lot of oil in the reservoir: typically, the recovery would be less than one-third of the oil initially in place.

It came to be realised that more of the oil could be recovered if the effects of the pressure drop could be delayed. One option is to install a pump near the bottom of the hole. That option is often applied, though it involves some challenging engineering, because the pump has to be small in diameter (so that it can go down the hole) and reliable (because production has to halt if the pump needs to be recovered for maintenance), and the pump has to be supplied with electric power.

Another option is to maintain pressure by drilling a second well and injecting water under pressure into that well. The water injected drives the oil towards the first production well (Figure 6.11(a)). Preferably, the boundary between the oil and the water remains nearly vertical, as in Figure 6.11(a), so that the displacement of the oil by the water is piston-like. It can instead happen that the boundary front develops irregular fingers, as in Figure 6.11(b), or reaches further under the oil than above it. When a finger reaches the production well, the well begins to produce water instead of oil ("watering out"): that is about to happen in Figure 6.11(b). Some of the oil can be bypassed by the fingers and left behind, and that oil is essentially unrecoverable, unless the quantity of oil can be determined confidently and it is judged worthwhile to drill a third well or more.

Modern thinking is that the best strategy is to maintain the pressure in the reservoir close to its initial value for as long as possible, by starting

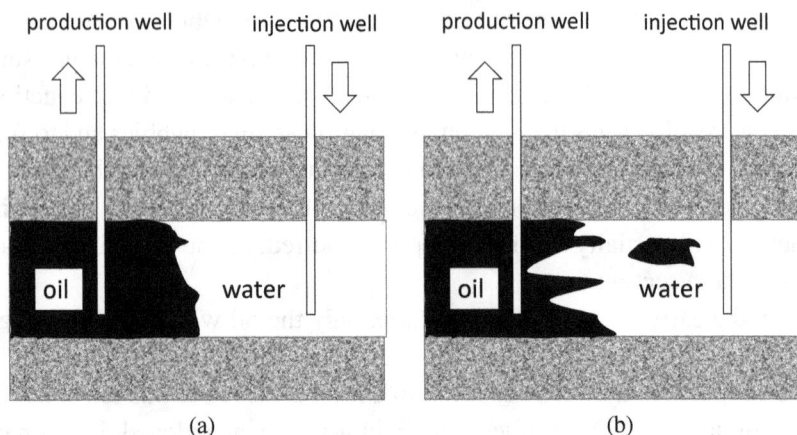

Figure 6.11   Water injection (a) piston-like displacement (b) fingering and bypassing

water injection early in the field life. The top of the oil remains near its original level, and does not move downward. The level of the bottom of the oil rises. That strategy also minimises the release of dissolved gas from the oil, and that gas too can partially block the flow of oil.

6.7 Well Spacing

Almost all fields need more than one well, and in an oilfield there are water injection wells in addition to producers. That raises the question of where the wells should be located. If there are many wells close together, the efficiency of producing the oil or gas is high, but so is the cost of the wells. If the wells are fewer and further apart, they cost less but some of the production is bypassed and left behind.

The relative mobility of water and oil is determined by the influences described in section 6.6. In some instances, it is primarily governed by gravity and the density difference, and in others by the ability of the fluids to move in and out of pore spaces, and by the intrinsic variability of the geology.

In featureless terrain on land, the wells can be located arbitrarily. They are often placed in standard patterns. A five-spot pattern has the producing wells located on a square grid, and an injection well at the centre of each of the squares, so there are equal numbers of producers and injectors. A direct line drive pattern has a line of producers at a regular spacing, a parallel line of injectors, a second line of producers, and so on. There are other standard patterns.

Offshore, there are more constraints. Each well is much more costly, particularly if it is in deep water and remote from a platform, and still more so if the well or a group of wells needs a platform of its own.

6.8 Gas Production from Hydrates

Gas hydrates are described in [2.5]. They are a huge potential source of energy for the future, but production is still some way ahead, though small-scale tests have been done in the Nankai Trough in Japan and in Siberia (Too *et al.*, 2015). In the present state of the petroleum industry, attention understandably focusses on easier short-term targets such as shale gas.

Start by looking at the stability diagram (Figure 2.4). A hydrate can only remain solid if the temperature remains low and the pressure remains low. It can be dissociated by lowering the pressure or increasing the temperature. A complication is that dissociation is endothermic, so that heat has to be supplied if dissociation is to continue. If the dissociation is started by lowering the pressure across the stability boundary, and if heat is not supplied, the hydrate temperature drops and the hydrate restabilises. Research has shown that the best option is to heat and depressurise at the same time (Falser *et al.*, 2012).

Heating and depressurisation can usefully be combined with hydraulic fracturing (Too *et al.* 2015).

References

Bame, D. and Fehler, M. Observations of long-period earthquakes accompanying hydraulic fracturing, *Geophysical Research Letters*, **13**, 149–152 (1986).

Begley, S. and McAlister, E. Earthquakes may trigger fracking tremors. *ABC Science*.

Falser, S., Uchida, S., Palmer, A.C., Soga, K. and Tan, T.S. Increased gas production from hydrates by combining depressurization and heating of the wellbore. *Energy and Fuels*, **26**, 6259–6267 (2012).

Falser, S., Palmer, A.C., Tan, T.S. and Loh, M. Testing methane hydrate saturated soil using a line dissociation apparatus. *ASTM Geotechnical Testing Journal* **35** (5) GTJ10431 (2012).

Gallegos, T.J. and Varela, B.A. Trends in hydraulic fracturing. *United States Geological Survey Scientific Investigation Report*, 2014-5131 (2014).

Hunt, J.M. Petroleum geochemistry and geology. W.H. Freeman and Company, New York (1996).

Kim, W-Y. Induced seismicity associated with fluid injection into a deep well in Youngstown, Ohio. *Journal of Geophysical Research: Solid Earth*, **118**, 1–13 (2013).

Mair, R.J. Shale gas extraction in the UK: a review of hydraulic fracturing. The Royal Society, London (2012), downloadable at https://royalsociety.org/~/media/policy/projects/shale-gas-extraction/2012-6-28-shale-gas.

Maitland, G.C. Oil and gas production. *Current Opinion in Colloid & Interface Science*, **5**, 301–311 (2000)

Reisberg, J. Surfactants for oil recovery by waterfloods. US patent 3348611 (1967).

Rubinstein, J.L. and Mahani, A.B. Myths and facts on wastewater injection, enhanced oil recovery and induced seismicity. *Seismological Research Letters*, **86**, 1–8 (2015)

Too, J.L., Falser, S., Linga, P., Boo, C.K., Cheng, A. and Palmer, A.C. An experimental method to determine the fracture toughness of brittle and heterogenous material by hydraulic fracturing. SPE-177014 SPE Asia Unconventional Resources conference, Brisbane (2015).

United States Geological Survey. How is hydraulic fracturing related to earthquakes and tremors? www.usgs.gov/faq/categories/10132/38830 (2015)

Chapter Seven

Producing Petroleum Under the Sea

7.1 Introduction

Understandably, the industry looked first in shallow water, in locations where petroleum had been known onshore and it was reasonable to expect that a favourable geology might extend under the sea. The early political and technological history is described by Feldman and Lagers (1997), and section 7.2 of this chapter summarises parts of their much more detailed research.

If the petroleum is to be produced, it has to be found. The first step is to learn as much as possible by exploration techniques, by integrating geological knowledge of the area with acoustic and chemical surveys. If an area looks promising, somebody has to decide to spend much more money on exploratory drilling. Drilling is described in Chapter 5. Exploration is by its nature uncertain, and the technology of exploration drilling is not the same as the technology used to produce.

7.2 Early History

Veldman and Lagers (1997) give a thoughtful and instructive account of the early history of drilling and production under the sea, an inspiration to engineers.

Summerland is close to Santa Barbara in California. It has natural tar seeps on the beach, and the explorer Cabrillo in 1542 reported that the local Chumash people used the tar as a waterproofing. In 1883 H.L. Williams bought land, initially with the notion of setting up a spiritualist colony, but soon afterwards he discovered oil, announced a few wells in 1887, and by 1895 had 28 wells. In 1897 he built a pier out into the sea, installed a cable-tool drilling rig, and produced oil that was of higher quality than the oil found onshore. By 1889 he had ten piers, one of them 400 m long ending in 10 m of water. Production was very slow: even the best well only produced 75 bbl/day, production offshore peaked before 1910, and soon the field was abandoned, though production onshore continued much longer. It is reported that oil is still seeping under water (Coastal Care, 2011). There is an extensive natural seep field nearby at Coal Oil Point (Leifer *et al.* 2010).

Similar production techniques were applied elsewhere along the California coast and in Lake Erie (Feldman and Lagers, 1997). In one relevant initiative, Rio Grande oil suspected that the Ellwood field near Goleta California extended under the sea, bought land onshore, and reached the undersea oil by directional drilling from land.

Natural gas seeps had been observed in Lake Caddo, on the Texas/ Louisiana border north-west of Shreveport, and gas was found in a water well for an ice factory. At the time there was little interest in gas. Oil was drilled for, but in 1905 found gas instead. That discovery was followed by a catastrophic series of fires and explosions and a huge waste of gas, leading to a federal inquiry in which a Geological Survey geologist spoke of

'the most flagrant abuse of natural wealth yet recorded in this industry'

(Veldman, 1997). Eventually a system was created. A platform was built of timber piles connected by diagonal beams. A 55 m length of nominal 10-inch (273 mm) pipe formed a conductor. Each platform carried its own equipment, and the oil was brought to gathering stations by nominal 3-inch (88 mm) pipelines. That scheme can be seen as the ancestor of today's offshore fixed platform.

The next offshore developments were in Lake Maracaibo in Venezuela, where a large oil field was discovered in 1922. Shell and Creole started with wooden platforms like those used in Lake Caddo, but found that they were rapidly destroyed by teredo shipworms. They moved forward to concrete piles and then in 1934 to steel piles, prompted by

'...the financial and economic advantages of steel over concrete'
'We did not desire to follow this system {of concrete piles} which required a great capital outlay for a concrete yard, and in our view it was not suitable for the greater depths of water' (Jansen and Kranendonk, 1950)

That was a bolder development, in which the industry was confident enough to free itself from the precedents of oil development on land, and could begin to make innovative moves on its own.

7.3 Mobile Submersible Drilling Rigs

The next idea was a drilling rig that could be moved from place to place without being dismantled. A key figure was Louis Giliasso: in 1928 he conceived and patented a floating vessel carrying a derrick, which could be towed to the desired position, sunk to form a fixed foundation, and later refloated and moved to another site. Working for the Texas Company, the forerunner of Texaco, another engineer McBride came up with the same idea, agreed to start building a vessel, found Giliasso — by that time running a bar in Panama — and secured the patent rights (Veldman and Lagers, 1997). The vessel was called after Giliasso, and was built from two barges, linked but with a space between them. It supported a derrick, the drilling equipment and a boiler to generate power. Two rows of piles fixed the vessel and protected the wellhead, and the vessel was sunk in position by controlled flooding. At the completion of drilling, it could be refloated.

The Giliasso was first deployed in 1933 on Lake Pelto. In the succeeding years, the offshore industry began to take off, particularly after the perceived shortage of US oil that worried US politicians concerned about another war. Many ideas were put forward. A persistent difficulty was that it was difficult to ballast a structure down to the seabed without it becoming unstable while it is still floating, and that difficulty increases with water depth. One of the most attractive designs was the 'bottle' submersible, where the platform rested on four columns, one at each corner, linked by a truss. The largest was the Kerr McGee Rig 54, which drilled in 50 m of water.

7.4 Jack-up Platforms

A jack-up platform can be thought of as an extension of the mobile submersible structure discussed in section 7.3. It derives from DeLong docks, which were used in the Second World War when ports could not be reached. The jack-up was developed in something like its modern form by Robert LeTourneau (1889–1969), a remarkably creative and influential engineer and manufacturer, owner of 300 patents and credited with a host of developments in earth-moving. Jack-ups are primarily applied in exploration drilling, but they can also be applied to production, as well as to other technologies such as offshore wind turbine installation

The hull is buoyant. The hull carries three vertical legs, and a rack-and-pinion system can move the legs up and down relative to the hull. The hull is towed to site with the legs in the up position. It lowers the legs one by one to the seabed, and applies a preload to seat the leg firmly. The lifting system can then lift the hull clear of the water, and the hull is a fixed base for the drilling operations. The legs are relatively small compared to the hull, and so once the hull has been lifted the wave forces are much smaller than they would be on a complete hull. Once the operations at one location are complete, the hull is lowered back into the water, and can be towed afloat to the next location.

Most jack-ups have three vertical legs that can be moved independently, but alternative options are to slope the legs outward, or to connect them through a pad which rests on the seabed and spreads out the load, or to have more legs.

7.5 Fixed Platforms

The next development was a bottom-founded fixed platform not intended to be moved. In 1938 Superior-Pure built the Creole platform in a 4 m depth 2.4 km from shore. However, the offshore industry often, though arguably mistakenly, dates its history from the first platform out of sight of land, the Ship Shoal block 32 Kermac 16, built in 1947 in 6 m of water 17 km from shore. In the words of its owners, Kerr-McGee:

> "we decided to move to areas where the really potential prolific production might be — salt domes — the good ones on land were gone, but we could find ... the real class-one type salt dome prospect"

Development proceeded rapidly, and the fixed platform became the workhorse of the offshore petroleum industry. Most platforms were open truss frameworks, fabricated from steel tubes and held to the seabed by

Figure 7.1   Jack-up by courtesy of Keppel Offshore and Marine

piles driven through pile guides attached to the legs, an option made easier by a parallel development of high-performance pile-driving.

There are many thousands of platforms of that type. They are usually constructed by building them on shore, dragging them from the fabrication site on to a barge, towing the barge to site, and tilting the barge until the platform slides off the barge into the water, where it floats with its feet above the seabed. The platform position is then adjusted and checked, and it is ballasted down to the seabed. The feet of the legs carry broad 'mud-mats' so that they do not sink too far into soft seabed. The piles are then driven to hold the platform in place. The length of the piles depends on the geotechnical conditions, but can be more than 100 m.

Figure 7.2 is an example. Looking from the west, it shows the ill-fated Piper Alpha platform in the North Sea, the site of a disastrous explosion and fire in 1988. The lowest level of the topsides is composed of four modules, from south to north drilling in module A, separation of gas and

Figure 7.2   Piper Alpha

oil in module B, gas compression in module C and the control room and various utilities in module D. The next level up has storage and drilling mud preparation, and above that accommodation and finally a helideck. The flare on the south side makes it possible to burn unwanted gas and oil (though routine flaring is nowadays not practiced). The platform has a crane to unload supply vessels. People travel to and from the platform by helicopter.

Figure 7.3 shows the Bullwinkle platform installed in 1988 in the Gulf of Mexico, the world's tallest pile-supported fixed steel platform, a record that is likely to stand for some time, in part because of the technological developments described in section 7.6 below. It has an overall height of 529 m and stands in 412 m of water in block Green Canyon 155, some 260 km south-west of New Orleans. It serves as a production hub for several fields. Marshall (2007) has written a thoughtful account of some of the difficulties met with in the construction of steel platforms.

Figure 7.3   Bullwinkle platform during floatout

Another option is to hold the platform in place by its own weight rather than by piles. Most gravity platforms are made of concrete, but a few are steel. Figure 7.5 depicts the Troll gravity platform in 303 m in the Norwegian sector of the North Sea. It was started in 1991 and built in sheltered water, and constructed while afloat, upwards from a base using slip-forming technology. The topsides were connected to the base while the base was afloat and ballasted down. In 1996 the platform was then towed 200 km from the construction site north of Stavanger. It weighed 680,000 tonnes at that stage. It is the largest structure that human beings have ever moved across the Earth's surface.

An advantage of a gravity platform is that it includes space that can be used for oil storage. A concrete option is often selected for reasons that are partly political, so that the platform can be built in the country where it is going to be installed. In Norway and in Newfoundland, for example, there is no domestic steel industry but a competence to build large objects out of concrete.

Figure 7.4   Bullwinkle platform after installation and during connection of pipelines (photo by courtesy of Bob Brown)

Figure 7.5   Troll platform (photo by courtesy of Harald Pettersen /Statoil)

7.6 Floating Systems

Rather than having a production structure firmly attached to the sea floor, an obvious option is to make the structure float. That option becomes particularly attractive in deep water, where the weight of a fixed structure would have to be carried a long way down to the seabed, the structural design would be heavily influenced by the wave and current forces that act near the top, and the structure would inevitably turn out to be somewhat flexible.

The produced fluids have to be exported, either by a shuttle tankers that go back and forth between the drillship and a shore refinery, or by export pipelines on the seabed. Even though the floating vessel is anchored, its anchoring system is flexible enough for the ship to move under waves and currents. The pipeline is stationary, and so the ship and the pipeline have to be connected by a riser that allows some relative movement.

Several alternatives have been developed, and there is much controversy about the best option in any particular combination of petroleum characteristics, marine environment, geotechnics, the need for separation, existing platforms and pipelines in the area, and distance to market.

A simple option is the self-propelled drillship. A vertical riser carries the fluids up from the formation to the vessel. Some separation can be carried out on the vessel, and the fluids can then be exported.

A second option is a floating vessel that has a different shape. Once it has been towed to site and anchored, it will not be moved again until it has completed its task at that site and needs to be moved to a new field. Towing resistance is then of little significance, but it becomes important to limit motions in rough seas, because if the sea gets rough it will be difficult and impracticably time-consuming to disconnect all the risers that connect the vessel to the seabed.

A configuration that accomplishes that requirement is the semi-submersible. The concept can be traced back to Edward Armstrong (1876–1955), who in 1922 proposed a 'seadrome' anchored in mid-ocean on which aircraft could land. An aircraft with limited range could then fly from one seadrome to the next and continue across the Atlantic. The runway deck was to be 370 m long and 60 m wide, held 20 m above the surface on slender tubular columns supported by buoyancy tanks at a depth of 30 m. Waves would pass underneath the deck. Model tests were carried out in a wave tank. The development of long-range aircraft meant that the seadrome was never built.

The application of the semi-submersible principle to offshore petroleum was originally put forward by Gross as early as 1946 (Veldman and Lagers, 1997). Since most of the buoyancy is well below the surface and the columns are quite slender, the effect is to reduce the forces induced by waves and to raise the natural period of roll above 20 s. The wave spectrum of a typical storm seastate has little energy at such long periods, and the roll movements are therefore small. The configuration responds to shifts of weight more than a conventional ship-shaped or rectangular hull does, but a ballasting system can compensate for that. A semi-submersible could in principle be self-propelled, but is normally towed into position. The semi-submersible scheme is often applied to other kinds of offshore construction vessels such as pipelaying barges.

A third option is a tension-leg platform (TLP). A floating hull is held in place by near-vertical tendons running down to anchors. The tendons are in tension, accomplished by ballasting down the hull, attaching the tendons, and then deballasting, so that the hull rises and puts the tendons into tension. There is almost no vertical movement of the TLP, and so the

Figure 7.6   Floating production system

Figure 7.7   Snorre tension-leg platform (photo by courtesy of Rune Johansen / Statoil)

wellheads can be on the TLP rather than on the seabed, which makes it easier to drill and complete the well, and to intervene if changes are needed. Horizontal movements are larger, but when the hull moves sideways, the horizontal components of the tendon tensions pull the hull back.

The first TLP was the Conoco Hutton TLP deployed in the UK sector of the North Sea in 1980. It has been followed by many other installations, in the Gulf of Mexico, Malaysia and Norway, including the Snorre TLP illustrated above. The greatest depth in which a TLP has been deployed is 1425 m for the Magnolia TLP. In very deep water, the weight of the tendons becomes a design problem.

Another option is a SPAR, which has the form of a large-diameter long cylinder, floating vertically and carrying the topsides. The name is said to derive from anchored logs that floated upright and marked navigation channels. A SPAR is held in position by mooring lines radiating outward. In the classic SPAR the whole hull is a cylinder, and in a truss SPAR the top is a cylinder and below it is a truss structure terminating in a ballast tank. The SPAR may carry helical strakes to suppress vortex-excited oscillations. Risers may run through a vertical well along the axis

Figure 7.8   Oryx SPAR (photo reproduced by permission from Offshore Technology conference, full citation in references)

of the cylinder, or they may be attached to the outside. The first SPAR was the Brent platform installed in the North Sea in 1976, though it was designed for storage and unloading rather than production. It was followed by many others (Vardeman *et al.*, 1997), the deepest the Perdido platform in 2440 m in the Gulf of Mexico.

It will be clear that there is no simple answer to the question of which floating production scheme is 'best'. Many factors have to be considered for each application, and regulatory authorities, company and personal preferences play a part.

7.7 Seabed Systems

All the systems described above have a structure that reaches the surface of the sea, either floating or coming up from the bottom. The sea surface is plainly a difficult place to locate anything unless it is essential to do so, because it often has large waves.

That suggests that a different option might be to locate the whole production system at the seabed, and to have it unmanned. We know from the developments of the past fifty years and more that almost every technological need can be met by machines, better and more cheaply and efficiently than by human beings on site. In the case of space exploration, for instance, it is cogently argued that development efforts ought to concentrate on unmanned systems. The perceived need for manned space exploration can be seen as a wasteful and irrelevant diversion, and the requirement for manned systems is only to meet politicians' need to keep interested the voters who will ultimately provide the funds. A seabed system would almost certainly have to be unmanned, because a manned system would encounter difficulties with safety, life support, fire protection, air and supply.

In the case of petroleum production, there are already not-normally-manned production platforms and wells operated remotely at some distance from shore or from a mother platform. The public and the regulators might welcome unmanned seabed systems, particularly if they could be reassured about safety and pollution risk. The idea of unmanned seabed systems has been put forward before (see, for example, Palmer and Loth, 1987).

All this is still in the future. The difficulties are substantial, but it is not fanciful to forecast that in fifty years' time engineers will look back with amazement at the fact that the offshore industry stuck for so long with manned systems at the surface.

7.8 Moving Petroleum from Offshore

Petroleum produced offshore has to be transported to shore so that it can be refined and used.

One way of bringing the petroleum to shore is by a pipeline. The first undersea pipelines were constructed from England to France in 1944, to carry petroleum for the Allied armies. Those lines were quite small, 88.3 mm (nominal 3-inch), but the technology was soon developed further, and it is now possible to lay pipelines in depths up to 3000 m and pipe made up at sea   pipe made up on shore

S-lay   J-lay   reel   pull and tow
                                                 bottom pull

                                                 surface tow

                                                 mid-depth tow

                                                 bottom tow

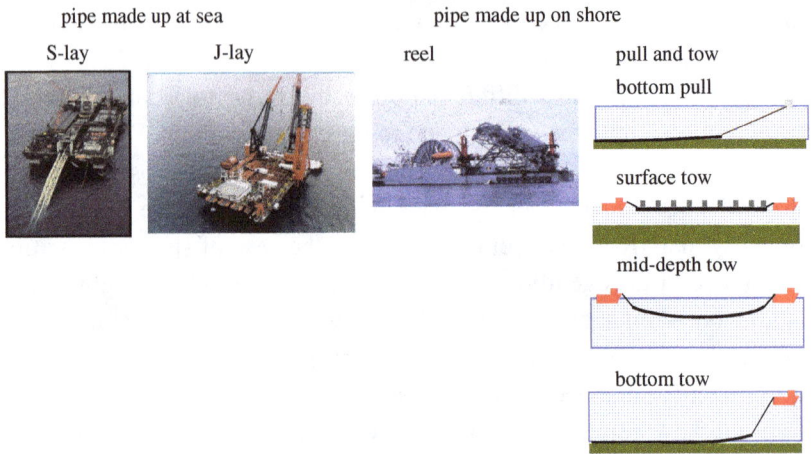

Figure 7.9   Alternative construction systems for underwater pipelines

in diameters up to 1219.2 mm (48 inches). If there were a demand for larger pipelines in deep water it could be met. The lines can carry oil or gas or both together.

Figure 7.9 illustrates how underwater pipelines can be constructed (Palmer and King, 2008). The two methods on the left are laybarge methods: lengths of pipe precoated with an anti-corrosion coating are brought to a barge and welded one by one to the end of the pipeline. The barge moves forward. The pipe moves through a series of welding stations, the weld is inspected and a coating is applied over the weld. The pipe leaves the barge over the stern and arcs down through the water until reaches the seabed. Tension controls the curvature of the suspended pipe. The pipeline is either near-horizontal on the barge ('S-lay') and curves round over a curved ramp called a stinger, or alternative on a steep ramp on the barge ('J-lay).

The methods towards the right of Figure 7.9 weld the pipe together onshore, carry it to site and lower it to the seabed. One option is to wind it onto a reel and then unwind it in place. Another option is to pull the pipeline into position, either by a fixed winch or by a floating tug, and either at the seabed or just below the surface or at an intermediate depth.

An underwater pipeline is protected against corrosion by a combination of an external coating and cathodic protection, which creates a small electrical voltage difference between the sea and the pipe. It sometimes

Figure 7.10   S-lay laybarge Lorelay (Allseas)

has an external coating of concrete, which increases its weight to stabilise it against currents and waves, and at the same times protects the anti-corrosion coating mechanically. Sometimes the line is trenched or buried a little way into the seabed. Trenching gives the pipeline some protection against damage by fishing gear and much reduces forces induced by currents and waves. Burial eliminates both those factors, provides thermal insulation, and controls buckling.

The other way of bringing petroleum to shore is by a shuttle tanker ship, an option in many ways simpler and more flexible than a pipeline, but often with a higher operations cost though a lower capital cost. That option is often applied to floating oil production systems. A tanker positions itself parallel to the floater and transfers oil through flexible hoses. The tanker option is most used for oil, but it can be extended to gas, either cooled and liquefied (Le Dévéhat, 2015) or highly compressed.

References

Coastal Care. http://coastalcare.org/2011/04/summerland-beachs-oil-seepage-mystery (2011)

Feldman, H. and Lagers, G. 50 years offshore. Foundation for Offshore Studies, Delft, Netherlands (1997).

Jansen, W.A. and Kranendonk, A. Nieuw type boortorenfundatie in het meer van Maracaibo. (New type of drilling tower foundation in Lake Maracaibo) De Ingenieur, **62**, 25 (1950).

Le Dévéhat, R. How to address offshore LNG transfer systems in terms of risks, safety and operation. Asian Natural Gas Infrastructure conference (2015).

Leifer, I., Kamerling, M.J., Luyendyk, B.P. and Wilson, D.S. Geologic control of natural marine hydrocarbon seep emissions, Coal Oil Point seep field, California. *Geo-Mar-Lett*, **30**, 331–338 (2010).

Marshall, P.W. Offshore technology: lessons learned the hard way. Lloyds Register Educational Trust lecture, National University of Singapore (2007).

Palmer, A.C. and King, R.A. Subsea pipeline engineering. Pennwell (2008).

Palmer, A.C. and Loth, W.D. A hybrid drilling system for deep water in the Arctic. *Journal of the Society for Underwater Technology*, **13**(2), 3–5 (1987).

UK Department of Energy. The public enquiry into the Piper Alpha Disaster. Her Majesty's Stationery Office, London (1990).

Vardeman, S, Richard, S. and McCandless, C.R. Neptune project: overview and project management. Proceedings, Offshore Technology Conference (1997).

Chapter Eight

Petroleum Accidents

8.1 Introduction

Petroleum is an immense activity. More than 10 million tonnes of oil are produced every day, and a similar amount of gas. It has to be found under the ground, drilled for, produced and refined, and transported and distributed to billions of users. It is scarcely to be expected that nothing will go wrong, just as it cannot be expected that a road transport network with 1.1 billion vehicles will never have accidents, or that billions of tonnes of coal can be burned without generating carbon dioxide pollution.

To say that is not to argue for complacency. "Any man's death diminishes me. No man is an island, entire in itself...". We need to take every possible step to minimise accidents, not only for reasons of morality but because accidents cause huge financial loss, consume time and energy, and damage the industry's reputation in the eyes of the citizen public. The hasty decisions that led up to the Macondo disaster saved at most a few days of rig time. The full costs are not yet known, and will perhaps never be known, but the purely financial loss is reported already to be more than 20 billion dollars, and some estimates are much higher still.

Pause to think of analogies. Exploration drilling is like drilling blindly into a high-pressure boiler, in the dark and two thousand metres away, not knowing exactly where the boiler is, just how thick it is, what it is made of or how large the pressure is. Petroleum is extremely flammable — that after all is why we want to produce it — and often toxic. In any petroleum activity, lots of it is present. If it gets loose, and if there is air around, it can easily burn or explode. If instead it is spilled on land, or worse still on water, it spreads out and is toxic to a wide range of life forms (including ourselves).

The most instructive practical approach is to examine mishaps exceedingly carefully, to understand what went wrong, and to record the lessons learned and make sure that they are publicised and acted upon. It is important to keep a sense of proportion and balance. A few months before the Macondo disaster, a friend of the author's was working for the oil company principally involved, and he was reprimanded for walking downstairs while reading a letter. At one level, one could argue that incident demonstrated an irrelevant concern for a small matter while much larger issues were overlooked. On the other hand, one could reply that a sensitivity to the possibility of accident ought to inform every aspect of our lives, and that small issues deserve awareness and action exactly as large ones do.

This chapter considers drilling first, then production onshore and offshore, and finally transportation. Each incident is referred to by the chapter number, the section number and the incident number. It is based on a small number of examples, but it cannot hope to be an exhaustive list or to go into great detail. Most of the incidents were the subject of extensive inquiry, and often of litigation or arbitration, and the reader interested in detail ought to refer to the reports. Inquiry reports are almost always thorough and careful.

The petroleum industry is much less good at reporting instructive mishaps than are the aerospace, railroad and pharmaceutical industries. There are several good books and papers about accidents: see, for example Perrow (1984), Bignell (1978), Pomeroy (2006), Faulkner (2003) and Duffey (2008) as well as inquiry reports on individual accidents. Several general observations emerge. Many and perhaps most accidents follow precursor incidents that ought to have been warnings, but either nobody noticed or nobody realised their significance and was concerned enough to take action. Human factors are extremely important.

In parallel there is a theory of reliability and quantitative risk analysis, but in the author's opinion it has little or no value or relevance to the petroleum engineer. A justification for that statement has been presented elsewhere (Palmer, 2012), but to be fair it must be added that many people do not agree.

8.2 Drilling Accidents

8.2.1 *Macondo*

The Macondo *Deepwater Horizon* disaster happened on April 20 2010. Eleven people were killed, five million barrels of oil were released into the Gulf of Mexico, and the result was the worst environmental catastrophe in the history of the US. Fitzsimmons (2011) has written a vigorous, hard-hitting and accessible short account. The incident is discussed at length in Lustgarten (2012). Bourne (2010) wrote a popular article, with a particular emphasis on environmental damage. The BOEMRE (Bureau of Ocean Energy Management, Regulation and Enforcement) of the US Department of the Interior wrote a careful and instructive report (2011), incidentally an instructive introduction to offshore drilling. There are many other reports, notably by BP, Transocean, and others. Unsurprisingly, the quality varies and the reports do not agree; for example, the BP report says that the well design did not contribute to the accident, but that opinion is not widely held. Several books have been written, and there are more books to come. Litigation will continue for years.

The well was first drilled from the *Marianas,* and later that rig was replaced by the *Deepwater Horizon,* a dynamically-positioned mobile offshore drilling unit (MODU) operated by Transocean, the world's largest offshore drilling contractor. The well was started ('spudded') on October 7 2008. It had been troublesome before:

> "At least three different well control events and multiple kicks occurred during operations at Macondo. On March 8 2010, it took the rig crew at least 30 minutes to detect a kick in the well... the drill pipe became stuck in the well and had to be cut free and bypassed" (BOEMRE, 2011)

The well was already five weeks behind schedule. The original estimated cost had been \$96M, but the actual cost was far over budget and \$154M had been allocated. The original target depth had been 6157 m (20200 feet) true vertical depth. If the pressure in a well is too high, the surrounding rock fractures; if the pressure is too low, fluids flow into the well. The difference between too high a pressure and too low is the 'drilling window'. In this instance, it was judged that the drilling window was too narrow, and it was decided to abandon the well temporarily at 5596 m (18360 feet), to install a production casing, and to return at some time in the future and make it a producing well.

There were two options for the production casing, a 'long string casing' that goes from the bottom to the top of the wellbore, and a 'liner casing' that is hung from inside the bottom of the previous casing string. A third option would have been to abandon the well without a production casing, but that would have increased costs by 10 to 15 M$, and was the option least preferred by BP.

The well engineering team 'Forward plan review' circulated earlier in April recommended the liner option, and against the long string option because

> "... cementing the long string was unlikely to be successful due to formation breakdown; using a long string would prevent BP from meeting regulatory requirements of 500 feet of cement above the top hydrocarbon zone; the long string would result in an open annulus to the wellhead, with hydrocarbon zones open to 9-5/8 inch seal assembly as the only open barrier..." (BOEMRE, 2011)

There was discussion in BP about the options. In the course of them, a BP engineer emailed another engineer and described Macondo as a "nightmare well which has everyone all over the place". In the end BP chose the long string option, and said that option would provide "the best economic case and well integrity case", whereas the liner option would add 7 to 10M$ to the cost of completing the well. However, the BOEMRE report said that the long string option was not a cause of the blowout.

The well had to be cemented, a routine procedure that isolates one zone from another and stops hydrocarbons from flowing up the annular space between the hole and the outside of the casing. The first step is to circulate mud to clean out the drill pipe and the casing. The mud is followed by the cement, and then by more mud to push the cement into position before it hardens. There is a valve device called a float collar; initially the valve allows flow in either direction, which is necessary when casing is run into the well, but then it is 'converted' and acts as a one-way valve to stop the flow from reversing. Cementing can be accompanied by problems: the cement may not harden properly (if it is contaminated) or it may leave unfilled spaces ('channelling'). Centralisers hold the casing in the centre of the hole, and a computer model that had been run a few days before said that channelling would not occur if 21 centralisers were used, but in the event only six were installed.

The float collar was supposed to convert when the pressure across it was between 2.8 and 4.8 MPa (400 to 700 psi). It appeared to convert only on the ninth try, when the pressure difference was very much higher at 21.66 MPa (3142 psi). The pressure required to circulate the well was much lower than had been predicted, but that was attributed to an inaccurate pressure gauge. The well site leader said "I'm afraid that we've blown something higher up in the casing joint". There is evidence that the float collar did not convert. A cement evaluation log test could have evaluated the quality of the cement, but the well team chose not to do it.

The integrity of the cement can be tested by increasing or reducing the pressure in the well. After the initial cement injection operations had been completed, and before the first negative pressure test, the cementing contractor carried out a flowback test at the cement pump. There was no backflow, and the test was thought to be a success. That conclusion was disputed by the inquiry later, and the anomalous negative pressure test results ought to have made it obvious that the backflow test could not have been successful. A 'positive pressure test' was successful. Then a 'negative pressure test' reduces the pressure and partially simulates what will happen when the well is abandoned. The drilling mud is displaced by seawater and a spacer fluid, both much lighter than mud, and so the well is underbalanced against the pressure in the formation. Loosely, the test invites the well to flow if it can. The negative pressure test produced anomalous results: the drill pipe pressure increased, an obvious sign that the well was possibly flowing. After discussion, a second negative pressure test was initiated, and started by closing a valve on the drill pipe, so that the pressure could not be monitored. There was no flow or pressure on the kill line, and the rig personnel decided that the test had been successful. The anomalous pressures were explained away as a 'bladder effect'.

The BOEMRE report concluded that hydrocarbons flowed from the reservoir, through the casing shoe (the lowest section of casing) and into the well and up the riser. This could only have happened if the cement in the shoe track failed. That might have been because mud contaminated the cement in the shoe track, or because the cement changed places with lighter drilling fluid, or because some of the cement flowed into the surrounding formation. The float collar may not have converted, but there could have been other reasons such as defective cement. A contributing factor was that the casing was set in a sand/shale zone close to an interval containing hydrocarbons.

Gas and oil flowed up the riser, expanding as they went because of the reduction of pressure. The blowout preventer failed to work, probably because the pipe in it was off-centre, had buckled, and had become trapped outside the cutting surfaces, so that they could not fully close and seal the pipe. Something ignited the gas and oil, and it exploded and started a fire. Eleven people were killed by the explosion. Several people jumped into the sea and were rescued.

It is clear from all the accounts that the people on the rig were under cost and time pressure, and that both were factors in their decisions to take short-cuts and omit tests. It can only have been managers on shore who applied that time pressure. Again and again, BP did not apply its own standard procedures. To some extent, the same applied to some of the contractors working for BP.

Human factors played a part, as they often do, for example in other accidents such as Aberfan. There were serious conflicts and lack of communication between managers, and the BOEMRE report includes revealing quotations. For example, an email three days below the blowout

"X, over the past four days there has been so many last minute changes to the operation the WSLs [Well Site Leaders] have finally come to their wits end. The quote is 'flying by the seat of our pants'. Moreover, we have made a special boat or helicopter run every day. Everybody wants to do the right thing, but, this huge level of paranoia from engineering leadership is driving chaos. This operation is not Thunderhorse. Y has called me numerous times trying to make sense of all the insanity. Last night's emergency evolved around the 30 bbls of cement spacer behind the top plug and how it would affect any bond logging (I do not agree with putting the spacer above the plug to begin with). This morning Y called me and asked my advice about exploring opportunities both inside and outside of the company. What is my authority? With the separation of engineering and operation, I do not know what I can and can't do. The operation is not going to succeed if we continue in this manner."

and a later email about the decision to use only six centralisers instead of 21

"But who cares, it's done, end of story, we'll probably be fine and we'll get a good cement job. I would rather have to squeeze than get stuck above the WH. So Z is right on the risk/reward equation"

The men called here X Y and Z are not identified by name, because there is no intention to single them out for special criticism; none of them was lost in the disaster.

Among the BOEMRE conclusions were that

"Many of the decisions leading up to the *Deepwater Horizon* blowout — including the timing of the installation of the lockdown sleeve, the conducting of multiple operations during mud displacement, and the use of lost circulation pills as spacer lowered the costs of the well and increased operating risks. These decisions were not subject to a formal risk assessment. BP's cost or time saving decisions without considering contingencies or mitigation were contributing causes of the Macondo blowout."

"Multiple decisions (the number of centralisers run, the decision not to run a cement evaluation, the decision not to circulate a full bottoms-up, and others) were in direct contradiction with the DWOP guidance to keep risk as low as reasonably practical. BP's failure to ensure all risks associated with operations on the *Deepwater Horizon* were as low as reasonably practical was a contributing cause of the Macondo blowout."

Fitzsimmons points out that only four months earlier there had been an 'eerily similar near miss' on the *Sedco 711* semi-submersible while it was working on the Bardolino field. Gas entered a riser while a well was being displaced with water during a completion operation.

"As at Macondo, the rig crew had already run a negative pressure test on the lone physical barrier between the pay zone and the rig, and had declared the result a success. The tested barrier nevertheless failed during displacement, resulting in an influx of hydrocarbons. Mud spewed onto the rig floor — but fortunately the crew was able to shut in the well before a blowout occurred."

8.2.2 *Lapindo*

A second serious and destructive event happened when the PT Lapindo Brantas exploration well Banjar Panji 1 in Indonesia targeted gas in the Kujung Formation carbonates. At a depth of 2834 m water, steam and gas erupted 200 m to the south-west. The well kicked and became a mud volcano, which grew and released enormous volumes of hot mud, at one time 180,000 m³/day. It came to flood 13000 ha, and destroyed 16 villages and displaced 50,000 people.

The general opinion is that the cause of the mud volcano was hydrofracturing of the formation, which was not protected by a casing. An alternative opinion is that the cause was reactivation of a fault by a magnitude 6.3 earthquake that had occurred two days before, 250 km to the south-west.

8.2.3 *Sea Gem*

The BP *Sea Gem* had made the first discovery of gas in the UK sector of the southern North Sea in 1965, in what is now the West Sole field. The vessel was a primitive jack-up, built by adding ten legs to a work barge. Three months later that year, the failure occurred while the barge was being lowered before a move to another site. Two of the legs collapsed, because of a combination of fatigue, corrosion, large environmental forces in rough seas, and unequal loading on the legs (Kemp, 2012; Burke, 2013). The rig fell over, and later capsized. Thirteen men were lost. There was no stand-by vessel, but by good luck a merchant ship was passing and saved 19 of the crew.

Two of the jacks had failed five weeks earlier, which ought to have been a warning. In consequence of the disaster, the UK made an important step forward, the Mineral Workings (Offshore Installation) Act 1971 (Kemp, 2012), which brought in the concept of an Offshore Installation Manager, with the same responsibilities for safety as the master of a ship.

8.2.4 *Ocean Odyssey*

The ODECO Ocean Odyssey semi-submersible suffered from a blow-out in September 1988 while drilling well 22/30b-3 for Arco in the North Sea (Norwegian Petroleum Directorate, 2013). It had drilled to 16160 feet (4920 m), but then began to lose mud. Not much mud and barite were left, and the Arco representatives chose to pull the drill string out of the hole to restore circulation, a decision contested by other members of the drilling team. Pulling the drill string out of the hole tends to reduce pressure at the bottom, and so encourages the well to flow. The team lifted the drill bit about 3000 feet (900 m), and stopped to try to circulate. The casing pressure suddenly rose, and mud and gas flowed up to the drilling floor. The crew were ordered to the lifeboats. The OIM (Offshore Installation Manager) told a radio operator to leave his lifeboat and return to the radio room. An explosion occurred, and ten minutes later a choke hose failure released a large amount of gas and caused fires on the semi-submersible and on the sea. The radio operator died on the rig from smoke and fire: he was the only fatality.

A Fatal Accident Inquiry was held in the Sheriff's Court. It was critical of Arco and the OIM. Arco had failed to respond to the shut-in drill-pipe pressure, failed adequately to respond to the gas kick, and did not shut in the well when it was out of control. It also said

> "The death of Timothy Williams [the radio operator] might reasonably have been prevented if (i) the OIM had not ordered him from the life-boat to the radio room (ii) if the OIM, having ordered [him] back to the radio room, had countermanded that order when the rig was evacuated..."

Disagreement between ODECO and Arco led to the disaster. The teams failed to recognise the pressure build-up when they were trying to recirculate the well, presumably trying to reduce mud losses by reducing the density of the drilling fluid (as also happened at Macondo; section 8.2.1). It is possible that concern for costs contributed to unwise decisions (again as at Macondo). It would have been possible to case off the zone that was losing mud, and then to continue.

8.3 Onshore Production Accidents

8.3.1 *Texas City*

A hydrocarbon vapour explosion and fire at a refinery in Texas City, Texas killed 15 people, injured 170 and led to widespread damage. Youtube (2015) has an excellent summary of what happened. Lustgarten (2012) describes the sequence of events in detail.

An isomerisation process was designed to convert low-octane hydro-carbons into higher-octane hydrocarbons that could be blended into unleaded gasoline. It included a 50 m tower that separated the lightest components, pentane and hexane. The accident happened because a blow-down stack had overflowed after an overpressure protection system had operated because of overfilling. The hydrocarbon vapour did not disperse because it was heavier than air, and it was probably ignited by a pickup truck, whose engine had overspeeded because of the hydrocarbons in the air. The consequences were made worse by the siting close to the isomeri-sation unit of contractor trailers manned by office workers.

An inquiry concluded that the accident was contributed to by cost-cutting, a failure to invest in updating the refinery, inadequate communi-cations, repeated failure to implement agreed procedures, defective equipment (which gave the operators completely incorrect information about levels in the tower), foolish siting of the trailers and poor training. Reports were critical of the culture at the refinery:

> "...The working environment had eroded to one characterised by resistance to change and lacking of trust, motivation and a sense of purpose. Process safety, operations performance and systemic risk reduction priorities had not been set and consistently reinforced..." (cited by Lustgarten (2012))

The immediate cause was an outdated blowdown drum that ought to have been replaced by a flare.

There had been numerous warnings. A report two months earlier had identified several safety issues. One of its authors had said "We have never seen a site where the notion 'I could die today' was so real". Starting 14 years earlier, it had several times been proposed that the blowdown system be replaced. OSHA (Occupational Safety and Health Administration) had mandated a switch to a flare system, but after appeal by the refinery owners the instruction had been withdrawn. In another refinery ten years earlier, two storage tanks had blown up and killed five workers in a trailer nearby.

8.4 Offshore Production Accidents

8.4.1 *Piper Alpha*

The Piper Alpha disaster in the UK sector of the North Sea happened on 6 July 1988. 167 people died, 165 from the platform and two from rescue vessels. The platform was a total loss. A public inquiry (1990) was followed by years of litigation, first in the Court of Session in Edinburgh, then in the Court of Appeal, and ultimately the UK House of Lords (the final court in the UK) reached its decision in 2002. The inquiry made 106 recommendations, and the disaster had a lasting impact on regulation in the UK, though not all the lessons learned were applied elsewhere.

The Piper Alpha platform exported oil by pipeline to a terminal at Flotta in the Orkney Islands, and gas to mainland Scotland by way of pipelines and an intermediate platform MCP01. Figure 7.2 is a photo of the platform. It was connected by gas pipelines to two nearby platforms, Claymore and Tartan. On the platform, the gas was first separated from the oil, and then condensate, the heaviest fraction of the gas, flowed from a flash drum and booster pumps to two parallel injection pumps, A and B, and was reinjected into the oil export pipeline. Each injection pump had a pressure safety valve (PSV), whose function was to ensure that if the condensate injection line failed, the PSV would allow the flow to be diverted to a relief line, so that the pumps would not be overpressurised. Only one injection pump had to be in operation at any one time.

On the evening of the accident, injection pump A was not operating, and its pressure safety valve PSV504 had been removed for maintenance. Injection pump B was working, but it stopped, and the operators tried to restart it. The details of what happened next are not completely certain, because those operators did not survive. The inquiry concluded that having tried but failed to restart pump B, the operators started pump A. That pressurised the blind flange that had been fitted when PSV504 had been removed. The blind flange had not been installed properly (or perhaps was missing altogether), and 45 kg of condensate escaped. The condensate mixed with air, and was somehow ignited and exploded. The explosion spread though module C, and broke down the firewall between module C and the control room in the next module D. At the same time it broke down the firewall on the other side of module C, and fractured another condensate pipeline in module B. That condensate in turn escaped, and initiated another fire. The fire rapidly got worse. Most of the people on the platform went up to the accommodation module, thinking that helicopters would come to take them off, but it was already too late for that. The gas risers from the Claymore and Tartan pipelines fractured because the fire heated them from outside, and they in turn burst and fed the fire further.

The inquiry looked at procedures on the platform, and concluded that the permit-to-work system that was supposed to control maintenance operations in a safe way was in practice sloppily operated. In the words of one of the survivors

> "Everybody had their own idea of how the system should be applied and it sort of changed week to week and crew to crew"

Even if the permit-to-work system had failed, there would have been no condensate release if the blind flange in place of PSV504 has been correctly fitted. This illustrates something often observed: the accident results from a chain of events, each one not enough itself to lead to the accident, but together creating it.

There had been a precursor accident on the same platform a year earlier, when a permit-to-work had been issued for a replacement of a thrust bearing but in reality a motor was lifted and a fatal fall resulted. Major accidents are often found to have been preceded by precursor incidents that were much less serious but ought to have been a warning: the *Titanic* disaster in 1912 and the Aberfan tailings tip disaster in 1967 are among many examples.

The Piper Alpha disaster is described in several reports and papers (Public Inquiry, 1990), and repays study.

8.4.2 *Alexander L. Kielland*

The *Alexander L. Kielland* semi-submersible drilling rig collapsed and sank in 1980, and claimed 123 lives. At the time of the accident the vessel was not acting as a drilling vessel but was serving as a 'floatel' providing living accommodation for the production platform *Edda 2/7C* in the Norwegian sector of the North Sea. The sea was rough with 12 m waves and a 40-knot (74 m/s) wind.

The semi-submersible had five legs. Leg D was connected by a horizontal brace D-6 to a node half-way along horizontal brace between legs E and C. A non-load-bearing flange plate supported a hydrophone. The flange plate was connected to bracing D-6 by a 6 mm fillet weld. The fillet weld had fatigued, in part because of a poor profile, and subsequent investigation found lamellar tearing in the flange plate and cold cracks in a butt weld in D-6. It found traces of paint in the crack, so at least some of the crack had already been present when the rig was painted.

Bracing D-6 broke. Leg D then broke away from the rest of the structure, and the rig tilted through 30°. Five of the six anchor cables broke immediately. Twenty minutes later the sixth cable broke, and the rig capsized. Out of seven lifeboats, only five were launched and only one was released from the lowering cables. Eighty-nine men were saved. The standby vessel did not arrive for an hour.

This incident reminds us that the failure of components that might be thought too small to deserve notice can have catastrophic consequences. The inquiry pointed out that 14 minutes went by between the initial failure of leg D and the final capsize, and that more men could have been saved if there had been an effective command structure.

8.5 Transportation Accidents

8.5.1 *Floating Pipeline*

A pipeline in the North Sea was discovered to have floated off the seabed over a distance of more than 500 m and to a maximum height of 18 m, so that it came close to the surface. Happily, the line did not fracture and by good luck a ship did not hit it, so no great harm was done, though the pipe did of course have to be put back to the bottom and restabilised. The incident was caused by loss of concrete weight coating caused by movements induced by a severe storm. The concrete was not up to today's standards.

8.5.2 *Bromborough*

An oil pipeline leaked 150 m³ and polluted an estuary. Movements had damaged the external anti-corrosion coating, water had reached the steel pipeline and corroded it, and ultimately a corrosion pit had penetrated the pipe wall. The operators did not notice that a failure had occurred until it was reported by ships. The leak occurred because pipeline movements had damaged the external anti-corrosion coating and allowed water to corrode the pipeline (Southgate, 1990). The subsequent inquiry was critical of the absence of a leak-detection system.

8.5.3 *Lac-Megantic*

A train of oil tank cars derailed and burned near the centre of Lac-Megantic, Quebec. Forty-seven people were killed, 1.6 million gallons of oil were spilled and ultimately a \$200 M settlement was reached. The train had gone out of control after an engine fire cut the power to its air brakes. It is widely thought that the cars were too lightly constructed. The same type of tank car was involved in several incidents before and since.

8.5.4 *Prince William Sound, Alaska*

The *Exxon Valdez* oil tanker went aground on Bligh Reef in Prince William Sound, Alaska, and released 40,000 m³ of Alaskan crude, about one-fifth of its cargo (though some estimates put the amount higher). Several factors contributed: the tanker was under the control of the third mate, the captain was asleep below deck, the course had been changed to avoid ice, and the radar collision avoidance system was broken (and had been broken for more than a year). The oil drifted 750 km south-westward through the Sound and into the Gulf of Alaska beyond Kodiak Island (van Bernem and Lübbe, 1997). The coast is remote, biologically important and difficult to access. Between a hundred thousand and three hundred thousand birds died, together with many sea otters and whales. An extensive clean-up followed. The incident had a damaging effect on the oil company and on the industry generally.

8.5.5 *Bellingham*

A 406.4 mm (16 inch) pipeline owned by Olympic Pipe Line Company ruptured and released 900 m³ (237,000 gallons) of gasoline into a creek that flowed through a park in Bellingham, Washington, US (NTSB, 2002). The gasoline flowed 2.5 km along the creek, and ignited and burned. Two children died of burns, and a man died from a combination of asphyxia and drowning. Property damage was more than $45M.

The source of the rupture was a gouge (a loss of wall thickness created by a sharp object) near a point where the Olympic pipeline was crossed by a water pipeline constructed later by a contractor, IMCO. The gouge was one of a group of gouges and dents close to the point where the rupture occurred. An electrician working for a subcontractor to IMCO stated that he heard the pipeline being struck by a backhoe during the project. It had been agreed that the last 0.6 m (2 feet) of excavation over a pipeline should be by hand rather than machine, but that appeared not to have been followed. The electrician who had heard the impact on the pipeline told the subsequent investigation that he had been present when the IMCO personnel decided not to tell Olympic or its inspector about the damage. IMCO coated the area that had been struck with mastic, a dense tar-like material that acts as anti-corrosion coating, and then buried it. Another worker said that he recalled an impact on a different pipeline. After the accident, a metallurgical examination of the pipe close to the rupture found traces of an element used to harden the teeth of excavators.

A subsequent survey by an intelligent pig (a device that is driven along the pipeline and measures corrosion, geometric defects and the pipeline's position) observed defects in the same area. They were analysed by a method intended for corrosion defects rather than sharp gouge defects, and it was concluded that they were too small to need excavation for further inspection and repair.

The accident occurred when the operating pressure at the rupture location was unusually high, but not higher than the pressure the pipeline ought to have been able to withstand if it had been undamaged. In the 30 minutes before the rupture, the operators had been aware of problems with the SCADA (supervisory control and data acquisition system) which monitors flow through the pipeline. The system used two parallel computers, one as the primary system and the other as a backup.

A pipeline controller was working on the primary system to create new records of vibration data, and about 10 minutes later the logs began to show errors. The flow of gasoline was switched as planned from one delivery point to another. Pressure in the pipeline began to increase, but that had happened before and was not unexpected. The customary response was to start a second pump at another pump station, but the system failed to carry out the command. At the same time, a high discharge pressure alarm further along the line indicated a pressure of 10.0 MPa (1444 psi), and later a marginal increase to 10.4 MPa (1494 psi). It was at about that time that the rupture happened, but the operators took some time to realise that. Within the next hour the pipeline was first shut down and then restarted, until reports began to come in, first of a strong smell and then a fire.

A pipeline system has relief valves to protect it against excessively high pressures. The inquiry found that the valves had not operated. The inquiry also found that the SCADA system was too easily accessible, that the records it kept were rather sketchy, and that it had no virus protection.

8.6 Conclusions

In every instance described above, possibly with one or two exceptions, the accidents described here were *not* primarily the consequence of 'technical' engineering failures. Almost always, human factors were important, sometimes overwhelmingly so. Many of the failures were preceded by disagreements among the people involved. In many cases there were precursor incidents that ought to have been a warning, a common feature of accidents in other fields (Bignell, 1978).

The industry has a 'gung-ho' 'can do!' attitude. In most ways that is an excellent thing: it is linked to energy, decisiveness and a readiness to take investment risks. The contrast with boring caution and bureaucratic inertia is recognised and enjoyed by people who come from different industries. Sometimes, though, that attitude becomes over-confidence and complacency, and examples show up in several of the incidents. Someone who argues against immediate decisions and taking risks, and for cautious reflection, is told that he is told that he is a weakling out of step with the spirit of the industry, that he is wasting money and time, and he is marginalised and pushed aside by those who want to be admired for their decisiveness.

References

Bartlitt, F. Presidential oil spill commission releases report from Chief Counsel, Fred Bartlitt. Bartlitt Beck Herman Palenchar & Scott LLP (2011).

Bignell, V., Peters, G. and Pym, C. Catastrophic failures. Open University Press (1978).

BOEMRE (Bureau of Ocean Energy Management, Regulation and Enforcement, US Department of the Interior. Report regarding the causes of the April 20 2010 Macondo well blowout (2011).

Bourne, J.K. The deep dilemma. *National Geographic.* October, 40–53 (2010).

Burke, L. The Sea Gem: a story of material failure. Journal of Undergraduate Research and Scholarship, Memorial University of Newfoundland, paper PT-13 (2013)

Duffey, R.B. and Small, J.W. Managing risk: the human element. Wiley (2008).

Faulkner, D. Shipping safety: a matter of concern. *Journal of Marine Design and Operations*, **B5**, 37–56 (2003).

Fitzsimmons, I. Macondo and the Presidential Commission. Offshore Engineer, March, 31–39 (2011).

Kemp. A. The official history of North Sea oil and gas. Routledge (2012)

Lustgarten. A. Run to failure: BP and the making of the Deepwater Horizon disaster. W.W. Norton & Co. (2012).

National Commission on the BP Deepwater Horizon Oil Spill. Deep Water: the Gulf oil disaster and the future of offshore drilling: report to the President (2011).

Norwegian Petroleum Directorate. Ageing semi-submersibles: review of major accidents (2003).

NTSB (National Transportation Safety Board). Pipeline rupture and subsequent fire in Bellingham, Washington, June 10 1998. NTSB/PAR-02/02 (2002)

Palmer, A.C. 10^{-6} and all that: what do failure probabilities mean? *Journal of Pipeline Engineering*, **12** (4) 269–271 (2012); *Pipelines International* (15) 55–57 (2013).

Perrow, C. Normal accidents: living with high-risk technologies. Basic Books (1984).

Pomeroy, R.V. Perception and management of risk-dependence on people and systems. World Maritime Technology Conference (2006)

Public Inquiry into the Piper Alpha Disaster. Her Majesty's Stationery Office, London (1990).

Southgate, D.A. Investigation report of the hot oil pipeline failure at Bromborough on Saturday 19 August 1988. Her Majesty's Stationery Office, London (1990).

US Chemical Safety and Hazard Investigation Board. Investigation Report — refinery Fire and Explosion, BP Texas City March 23 2005.

Van Bernem, C. and Lübbe, T. Öl im Meer: Katastrophen und langfristige Belastungen (Oil in the sea: catastrophes and longterm impacts). Wissenschaftliche Buchgesellschaft (1997).

Youtube. www.youtube.com/watch?=VCcN45Qkb9A (2015).

Chapter Nine

The Future of Petroleum

9.1 Introduction

The industry is far from static, and there have been many surprises.

In the 1930s, nobody began to anticipate the scale of development that would occur. A distinguished petroleum engineer poured scorn on the prospects in Saudi Arabia, and famously declared that he would drink all the oil found there. Fifty years ago, a petroleum engineer who decided to move to Norway would have been judged crazy. Only ten years ago, students were taught that gas and oil could not be produced from impermeable shale formations, whereas nowadays shale gas and oil are productive and highly fashionable. It seems likely that they will enable the US to become a net exporter, with huge economic and geopolitical consequences. At the time of writing, the oil price has halved in three months, as it has done before. Daniel Yergin, an enormously distinguished energy guru who has authored several important books, writes in today's newspaper (2016) to say that the price will recover, but he is noticeably careful not to predict how long that will take and not to say that the price will not dip further.

It is almost certainly useless to try to forecast the future. Almost nobody predicted the collapse of the Soviet Union, or the Ebola epidemic, or the rise of terrorism as the primary current security concern. All the same, it is worthwhile and stimulating to think what the surprises might turn out to be. It might even be productive.

9.2 Technological Developments

9.2.1 *Fracking and Horizontal Drilling*

One major development has been the combination of horizontal drilling and fracturing ('fracking'), which together release oil and gas from low-permeability sedimentary rocks previously judged impossible to produce economically. That development was pioneered onshore in the USA, but we can reasonably expect it to extend offshore and to other countries. Groups hostile to the oil and gas industry have seized on fracking as something to object to, but we can expect resistance to die away as the industry communicates better. The word 'fracking' was perhaps an unfortunate choice, but it is too late to change.

9.2.2 *Heavy Oil*

The continuing success of fracking is bound to encourage a wider development of ways to release gas and oil from difficult formations. Similarly, and if the oil price is favourable, people will think more about heavy oil, with a gravity below 20 or a viscosity above 200 cP, and extra heavy oil with a gravity below 10 or a viscosity above 10,000 cP; definitions vary slightly. They are found in huge quantities in Alberta, Venezuela, California, Madagascar, Chad, Angola, Kuwait and some 20 other countries. Some estimates say that world reserves of heavy oil, extra-heavy oil and bitumen together add up to more than twice conventional oil reserves: this issue is discussed in more detail by Meyer and Affanasi (2003). Heavy oil often but not always contains a high proportion of sulphur, up to 4.5%. Heavy oil is generally thought to have been formed from lighter oil by biodegradation, in which bacteria have preferentially consumed the light fractions, or by water percolation which again has removed the light fractions.

Heavy oil is too viscous to be produced by conventional technology, and commands a much lower price. The discount for Canadian heavy oil, as measured by the WCS (Western Canada Select) price differential to WTI (West Texas Intermediate), averaged US$13.27/bbl (or 29%) in Q3/2015 as compared to US$11.59/bbl (or 20%) in Q2/2015 (Platts, 2015). BP (2013) concluded that heavy oil 'is unlikely ever to be more economic than light oil'. That does depend on how you interpret 'economic': does it mean that a light oil development will always be more attractive financially than a heavy oil one (no matter where it is or how large it is or what the acreage cost), or that light oil will always generate more profit per barrel? The balance of advantage might of course shift in response to technical development.

A related form of petroleum is found in bituminous oil sands, found in vast quantity near Fort McMurray in northern Alberta Canada and near the Orinoco river in Venezuela. Oil sands are loose sands saturated with dense bitumen and extra-heavy oil. They were formerly called tar sands, but that term was thought to be environmentally provocative. The Canadian oil sands are near the surface, and the bitumen is extremely viscous. The Orinoco sands are much deeper, but the oil is less viscous, so that it is sometimes possible to extract it by conventional techniques.

Some heavy and extra-heavy oil development has already taken place. Oil has been produced by surface mining from the Canadian sands since 1967, and the mining operation is now on a huge scale. It was intended to export that oil to the USA by a new pipeline, but that controversial option is indefinitely delayed by political action in response to environmental campaigns. Venezuela has produced 9°API oil in relatively small 100,000 b/d quantities, and exported it by tanker as 'Orimulsion', an emulsion of 70% oil and 30% water. That development is beset by politics, and has been scaled down.

CHOPS (Cold Heavy Oil Production with Sand) draws down the well rapidly at first, deliberately so that sand flows into the well with the oil, the opposite of what is usually wanted. The flow of sand creates wormhole channels and increases the area that oil can flow into. The sand has to be separated at the surface. SAGD (Steam Assisted Gravity Drainage) drills two horizontal wells, one above the other. Steam flows into the upper well, and heats the formation, so that the oil becomes less viscous. The oil flows into the lower well. A third possibility is to heat the oil down a well, ignite it, and use the steam generated from associated water to heat the oil and drive it towards a second well. One version of this idea is THAI (Toe to Heel Air Injection), a form of fireflood, which has air injection down a vertical well to fire within the formation, and produces from a horizontal well.

The oil has to be transported to where it can be used. One option is to refine it on site. Another is to dilute it with lighter oil or mix it with water, so that it can flow along a conventional pipeline. A third option is to transfer heat to the pipeline, using an existing heat tracing technology developed for other applications, but a problem is that if the heat tracing stops working the oil cools down and flow is hard to restart.

9.2.3 *Deep Water*

It is already possible to produce oil from reservoirs under 3000 m of water, and to carry out seismic exploration and exploratory drilling in still deeper water. If the attractive reservoirs are there, there seems no reason not to believe that production can be extended to any depth. It will be important to contain costs, and not to allow enthusiasm to persuade oneself that any hydrocarbon development will be economic, no matter how high the costs. Much can still be done to reduce costs by standardising designs and directing development to stable countries that combine a trained workforce with low prices.

9.2.4 *Unmanned Platforms*

Looking further ahead, current offshore developments depend on structures at the surface, whether platforms or floating production systems (FPS).

Almost always, those structures are manned. It is costly to support a person offshore: a support cost is many hundred dollars per day of work. Besides his pay, he (or she) needs to work on a rotation (so that he can go home to rest and see his family), he has to be transported to and from the platform or the FPS, he needs to be fed and have clean sheets on his bed, he needs quiet space to sleep, he needs some facility for exercise and relaxation, and he needs to be able to leave safely in case of mishap. All those things are expensive, and are likely to become even more expensive as the labour union movement extends its range and influence. That suggest a move to not-normally manned (NNM) platforms and FPS, a move that has been made easier by the development of reliable and immediate communications systems. The nomenclature varies with the degree of development of the philosophy: NNM is not-normally manned or normally-not-manned, NUI is normally unattended, and MMI is minimally manned.

Many operators have already adopted this approach, and is not fanciful to imagine than in not many years' time NNM will be the norm rather than the exception. One is reminded of the automated factory operated by a man and a dog: the dog is there to stop the man from touching anything, and the man is there to feed the dog.

The first NNM platform in the North Sea, Q8 in the Netherlands sector, went into service in 1986, and from the beginning was operated and monitored by telemetry to and from shore. Another example is the Woodside Angel platform on the North-west Shelf of Australia, a remotely powered and remotely operated gas dehydration platform that takes gas from three subsea wells, reported to be the most complex unmanned platform to date. It produced first gas in October 2008, two months after start-up, and attained NNM status seven months later. Operations and planned maintenance activities are carried out during a two or three day visit every six weeks. There is temporary accommodation for 20 people.

West Sole in the North Sea has two platforms A and B, constructed in the 1960s and fully manned until the early 1980s with 50 or 60 people on board. The number of people on board was progressively reduced, first to 20 and then to 12 (Edwards, 2013). At the time of writing, West Sole has five operational platforms, and two of them are unmanned, WB and WC. It is hoped that all the platforms can be made NNM. A final example is BP Hoton, a small gas development to the north-east of West Sole, completed in 2001, at a project delivery cost stated as $80 million, to be compared with $350 million for the traditional project approach.

In the UK sector of the North Sea, there are currently 80 oil platforms, of which 13 are unmanned, 20 floating oil installations, all manned, and 189 gas platforms, of which 135 are unmanned (https/www.gov.uk, 2014). There are also many NNM systems on shore. In all these cases, adoption of NNM can be seen as part of a wider exercise of value chain integration (Edwards, 2013).

9.2.5 *Facilities on the Seabed*

A further step is to put facilities in a better environment. After all, the surface of the sea is a bad place to put anything, unless there is no alternative. The surface is a hostile environment. There are often rough seas, and occasionally there are huge waves: the design maximum wave height in the North Sea and in the Gulf of Mexico is between 25 and 30 m, depending on the exact location. Waves as large as that exert very large forces, and are difficult to design for, while they make life difficult for anyone on a floating system. In the Arctic there are also sea ice and icebergs, which can be very large indeed.

An option is to put facilities on the sea bottom (Palmer and Loth, 1987). That moves from one challenging environment to another. Remote wells and manifolds on-bottom are nowadays routine. It would scarcely be practicable to have an on-bottom structure continuously manned. Half a century of routine deployment of nuclear submarines might suggest otherwise, and the technology is within our grasp, but the cost of continuously-manned seabed facilities would be very high. In contrast, the NNM (not-normally-manned) concepts described earlier in this section are progressing very rapidly, and complex on-bottom production installations appear technically feasible.

9.3 Economic and Political Developments

These are even more difficult to forecast.

An opinion widely held is expressed by ExxonMobil in its 2013 Summary Annual Report (2013) and the later press release (2014): the figures below are taken from the latter. It predicts a 35% increase in energy demand by 2040, accompanying 'significant growth in the global middle class, expansion of emerging economies and an additional 2 billion people in the world…'. It projects that carbon-based fuels will continue to meet about three-quarters of global energy demand through 2040, and says that that is

> '…consistent with all credible projections, including those made by the International Energy Agency. The outlook shows a shift towards lower-carbon fuels in the coming decades that, in combination with efficiency gains, will lead to a gradual decline in energy-related carbon dioxide emissions.
>
> 'Wind, solar and biofuels are expected to be the fastest-growing energy sources, increasing about 6% a year on average through 2040,

when they will be approaching 4% of global energy demand. Renewables in total will account for 1 % of energy demand in 2040. Nuclear energy, one of the fastest-growing energy sources, is expect to nearly double from 2010, with growth in the Asia-Pacific region, led by China, accounting for about 75% of the increase.'

If that is correct, we can expect that for the coming 25 years oil and gas, together with coal, will remain the primary source of energy for the world's population.

That opinion is of course bitterly contested. A significant group of people takes an opposite view, and sees the threat of climate change induced by carbon dioxide in the atmosphere as so serious that humanity needs to make radical changes in its energy sources, and needs as soon as possible to replace most if not all fossil fuels by renewables. Not everyone agrees that carbon dioxide is the source of climate change and that carbon-based fuels are to blame; for opposing views see, for example, Gore (2006), Carter (2010), Houghton (1994), and Lomborg (2001). Nuclear energy does not dump carbon dioxide into the atmosphere and is clearly an alternative, but has its own problems of potential pollution and waste, and of a population made frightened by the association with nuclear weapons; see, for example, Weart (2012), Brand (2009) and Kidd (2008). Some people are even more hostile to nuclear energy than they are to fossil fuels. Non-nuclear renewables are at an early stage of development, and thus far contribute very little, but they could potentially take over a significant share of the world's energy needs.

All this is of course against a background of the world's economy and politics. It is probable that there will be huge changes, as there have been in the past century, for example a worldwide epidemic or a nuclear war. It is sometimes argued that climate change is the most serious challenge that humanity will have to face. That fear is grossly exaggerated. If climate change is the worst thing that happens in the coming century, we shall have been incredibly lucky.

References

Baytex.http://www.baytexenergy.com/files/pdf/Operations/Q3%202015%20 Heavy%20Oil%20Pricing%20Update.pdf (2015).

BP.http://www.aoga.org/wp-content/uploads/2011/03/HRES-3.9.11-Lunch-Learn-BP-Heavy-Oil1.pdf (2013).

Brand, S. Whole earth discipline: an ecopragmatist manifesto. Viking (2009).

Carter, R.M. Climate: the counter consensus. Stacey International (2010).

Edwards, T. Integrated operations enabling new operational and project concepts. Ninth International conference on Integrated Operations in the Petroleum Industry, Trondheim, Norway (2013).

Exxon Mobil Summary Annual Report (2013).

Gore, A. An inconvenient truth: the planetary emergence of global warming and what we can do about it. Rodale (2006).

Houghton, J. Global warming: the complete briefing. Cambridge University Press (1994).

http://cdn.exxonmobil.com/ ~/media/Reports/Outlook%20For%20Energy/2015/US-2015-Outlook-for-Energy-press-release.pdf, ExxonMobil's outlook for energy sees global increase in future demand. (2014).

Kidd, S. Core issues: dissecting nuclear power today. Nuclear Engineering Special Publications (2008).

Lomborg, B. The skeptical environmentalist: measuring the real state of the world. Cambridge University Press (2001).

Meyer, R.F. and Affanasi, E.D. Heavy Oil and Natural Bitumen — Strategic Petroleum Resources. US Geological Survey factsheet 70–73 (2003).

Weart, S.R. The rise of nuclear fear. Harvard University Press (2012).

www.gov.uk/government/uploads/system/uploads/attachment_data/file/385346/Installations_December_2

Appendix
Units

Much of the oil industry uses non-metric units, whereas most of the world uses metric units for any serious purpose, usually but not invariably in the SI (Système Internationale) form. That situation is deeply unfortunate, but there is little or nothing that can be done about it now. Mixed units and confusion about units are frequent sources of mistakes, sometimes very costly ones. The retention of old units rather than metric units happened because of the leading position taken by the US industry in the early years, even though several petroleum regions were developed by Europeans and Asians rather than Americans. The International Standards Organisation (ISO) has produced a specification for SI (1981). The rule is that units named after people are given capital letters (e.g. Pa for Pascal, the SI unit of pressure) but m for metres. Multiples and submultiples are indicated by prefixes, k for thousands, M for millions, G for billions (one thousand millions) and m for one-thousandth, so that 1 GPa means 1,000,000,000 Pa and 1 mPa means 0.001 Pa.

In the units most used in the petroleum industry, oil volumes are measured in barrels, abbreviated b or bbl. One barrel is 42 US gallons (not the same as UK gallons), and is 0.158987 m^3. Gas volumes are measured in standard cubic feet, written scf and often pronounced 'scuff'. One standard cubic foot is $(0.3048 \text{ m})^3 = 0.0283168 \text{ m}^3$. Gas is usually measured by volume at standard conditions. The reference to standard conditions is important, because gas is compressible and expands in proportion to temperature measured from absolute zero. There is more than one set of standard conditions, and the most used are 60°F (15.55°C) and 1 atmosphere (101.325 kPa) pressure, but sometimes 15°C and 1 bar

(100 kPa); the difference is small. One scf of gas is rather a small volume, and multiples are Mscf Thousands), MMscf (millions), Bscf (billions) and Tscf (trillions).

SI measures length in metres m, and the multiples mm (millimetres, 10^{-3} m, 1 thousandth of a metre), microns μm (10^{-6} m, 1 millionth of a metre) and nanometres nm (10^{-9} m, 1 thousand-millionth of a metre). Those are the preferred sub-multiples, but other sub-multiples such as the centimetre (cm, 10^{-2} m) and Angstrom unit (10^{-10} m) are still in use, particularly in countries that adopted the metric system some time ago. In the American system, lengths are measured in inches or feet or miles. One inch is 0.0254 m exactly, because that is how the inch is defined. One foot is 12 inches and therefore 0.3048 m. Depths are measured in feet. Depths in the sea are usually in feet but sometimes in fathoms: one fathom is 6 feet. One mile is 5280 feet, which is 1.60934 km. At sea, lengths are often measured in nautical miles, which are different again: one International nautical mile is 1.852 m, but very slightly different definitions are used in some countries.

It is important to bear in mind that by convention some kinds of dimensions are cited as standard dimensions, but are not exact conversions. What is usually called a '10-inch pipe' does not have an outside diameter of 10 inches, but instead 10.75 inches.

In the metric system, SI makes a clear distinction between forces (measured in newtons N) and masses (measured in kilograms kg) , and is consistent in the sense that a force of 1 N applied to a mass of 1 kg generates an acceleration of 1 m/s². That consistency is missing in the non-metric and non-SI systems, another source of mistakes. In metric but not SI systems, force is often measure in kilograms force (kgf), in some countries called kilopond (kp); 1 kgf is 9.80665 N. In the American system, masses, forces and weights and are all measured in pounds, abbreviated lb (from the Latin *libra* for pound). One pound is 0.4535237 kg. One American ton is 2000 lb (907.047 kg), but in the UK a ton is 2240 lb (1015.89 kg), and many countries still use the metric tonne, which is 1000 kg = 2204.96 lb. The differences are significant.

In the SI system, pressure and stress are measured in Pa. One Pa is 1 N per square metre. Because 1 Pa is a very small pressure, the kiloPascal (kPa, 1000 N/m²) and megaPascal (MPa, 1,000,000 N/m²) are more commonly used. In the American system, pressure and stress are measured in pounds per square inch, written psi or lb/in. One psi is 6894.65 Pa = 6.89465 kPa.

In the SI system, temperature is measured in degrees Celsius (°C). On that scale, the freezing point of water is 0°C and the boiling point is 100°C. Degrees Celsius were formerly called degrees centigrade, but that usage is discouraged nowadays. Some of the English-speaking world still use degrees Fahrenheit (°F). On the Fahrenheit scale, the freezing point of water is 32°F and the boiling point of water is 212°F, both at standard atmospheric pressure. Absolute zero, −273.15°C, is −459.69°F.

References

ISO 1000–1981 Specification for SI units and recommendations for the use of their multiples and of certain other units.

ISO 31–1981. Specification for quantities, units and symbols.

Index

T

tension-leg platform, 98
Texas City, 114
THAI (Toe to Heel Air Injection), 125
thixotropy, 58
three-cone bit, 54
tidal power, 5
tool joints, 53
toolpusher, 61
total acid number, 19
total stress, 26
trap, 32, 34
tri-cone, 54
Troll platform, 96
two-way travel time, 45

U

unconformity, 34
units, 131

V

viscosity, 19

W

watering out, 85
water saturation, 78
Wegener, 27
weight on bit, 53
well spacing, 86
West Sole, 112, 127
whipstock, 62
wind energy, 4
Woodside Angel platform, 127
world primary energy
consumption, 1
Wytch Farm, 37, 42, 75